Spatial and Temporal Dynamics of Cyanotoxins and Their Relation to Other Water Quality Variables in Upper Klamath Lake, Oregon, 2007–09

By Sara L. Caldwell Eldridge, Tamara M. Wood, and Kathy R. Echols

Prepared in cooperation with the Bureau of Reclamation

Scientific Investigations Report 2012–5069

U.S. Department of the Interior
U.S. Geological Survey

U.S. Department of the Interior
KEN SALAZAR, Secretary

U.S. Geological Survey
Marcia K. McNutt, Director

U.S. Geological Survey, Reston, Virginia: 2012

Suggested citation:
Eldridge, S.L.C., Wood, T.M., and Echols, K.R., 2012, Spatial and temporal dynamics of cyanotoxins and their relation to other water quality variables in Upper Klamath Lake, Oregon, 2007–09: U.S. Geological Survey Scientific Investigations Report 2012–5069, 34 p.

Contents

Abstract...1
Background...1
    Study Area..3
    Purpose and Scope 3
Methods 5
    Water Quality Data Collection 5
    Cyanotoxin Analysis 7
    Data Analysis...7
Interannual Variability, 2007–09...8
    Chlorophyll *a* and Nutrients 8
    Dissolved Oxygen, pH, and Temperature..10
    Cell Associated Microcystins 12
    Dissolved Microcystins...12
Relations Between Microcystin Concentrations and Other Water Quality Variables 16
Seasonal Variability, 2009 18
    Chlorophyll *a* 18
    Total and Dissolved Nutrients 18
    Nutrient Ratios...21
    Particulate Carbon, Nitrogen, and Phosphorus...21
    Continuous Monitor and Meteorological Variables 23
Spatial Variability of Microcystin Concentrations, 2009..23
Relation Between *Microcystis aeruginosa* and *Aphanizomenon flos-aquae*.................................24
Implications for Juvenile Sucker Health...25
Summary and Conclusions 26
Acknowledgments...27
References Cited..27
Appendix A. Quality Control and Quality Assurance of Water Samples 33
Appendix B. Results of Microcystin Analysis 34

Figures

Figure 1. Map showing locations of meteorological station and cyanotoxin sampling sites, Upper Klamath Lake, Oregon, 2007–09 4

Figure 2. Graphs showing median concentrations of large (> 63 µm) particulate (cell associated) microcystins, dissolved microcystins, chlorophyll *a*, total nutrients (nitrogen and phosphorus), the ratio of total nitrogen to total phosphorus, dissolved nutrients, and the ratio of dissolved inorganic nitrogen to dissolved inorganic phosphorus in water samples from Upper Klamath Lake, Oregon, 2007–09 9

Figure 3. Graphs showing ratio of chlorophyll *a* to total phosphorus concentrations in water samples collected from Upper Klamath Lake, Oregon, 2008–09 10

Figure 4. Graphs showing median water temperature, pH, dissolved oxygen concentrations, and dissolved oxygen percent saturation at sites in Upper Klamath Lake, Oregon, 2007–09 11

Figure 5. Graphs showing percentage of microcystin concentrations representing each size fraction at sites in Upper Klamath Lake, Oregon, 2009 14

Figure 6. Graphs showing median concentrations of total microcystins and chlorophyll *a*, total phosphorus and total nitrogen, orthophosphate (dissolved inorganic phosphorus), and ammonia and nitrite plus nitrate (dissolved inorganic nitrogen) in Upper Klamath Lake, Oregon, 2009 19

Figure 7. Graphs showing median concentrations of total microcystins and the ratio of total nitrogen to total phosphorus and the ratio of dissolved inorganic nitrogen to dissolved inorganic phosphorus in Upper Klamath Lake, Oregon, 2009 21

Figure 8. Graphs showing median concentrations of total microcystins and total particulate nitrogen, total particulate phosphorus, total particulate carbon, and the ratio of total particulate nitrogen to total particulate phosphorus in Upper Klamath Lake, Oregon, 2009 22

Figure 9. Map showing relative median concentrations of microcystins in the dissolved and large (> 63 µm) particulate (cell associated) fractions at sites in Upper Klamath Lake, Oregon, 2009 24

Tables

Table 1. Summary of samples or data collected for analyses of cyanotoxin occurrence and other water quality parameters, Upper Klamath Lake, Oregon, 2007–09 6

Table 2. Median, range, and percentage of values below the method quantitation limit of microcystin concentrations in different size fractions collected between July 7 and September 14, Upper Klamath Lake, Oregon, 2007–09 13

Table 3. Median and range of large (>63 µm) particulate (cell associated) and dissolved microcystin concentrations at each sample site between July 7 and September 14, Upper Klamath Lake, Oregon, 2007–09 15

Table 4. Spearman rank order correlations (r) between microcystin concentrations and other environmental variables, Upper Klamath Lake, Oregon, 2007–09 17

Table 5. Spearman rank order correlations (r) between microcystin concentrations and other environmental variables, Upper Klamath Lake, Oregon, 2009 20

Conversion Factors and Datums

Conversion Factors

Multiply	By	To obtain
gram (g)	0.03527	ounce, avoirdupois (oz)
microgram (μg)	3.527×10^{-8}	ounce, avoirdupois (oz)
kilometer (km)	0.6214	mile (mi)
meter (m)	3.281	foot (ft)
nanometer (nm)	3.937×10^{-8}	inch (in.)
micrometer (μm)	3.937×10^{-5}	inch (in.)
millimeter (mm)	0.0394	inch (in.)
liter (L)	1.057	quart (qt)
milliliter (mL)	0.03382	fluid ounce (oz)
meter per second (m/s)	3.281	foot per second (ft/s)

Temperature in degrees Celsius (°C) can be converted to degrees Fahrenheit (°F) as follows:

$$°F=(1.8\times°C)+32.$$

Concentrations of chemical constituents in water are given either in milligrams per liter (mg/L), which is approximately equal to parts per million, or micrograms per liter (μg/L), which is approximately equal to parts per billion (ppb).

Datums

Horizontal coordinate information is referenced to the North American Datum of 1983 (NAD 83).

Elevation, as used in this report, refers to distance above the Upper Klamath Lake Vertical Datum (UKLVD), which is used by the Bureau of Reclamation for reporting the elevation of Upper Klamath Lake.

Spatial and Temporal Dynamics of Cyanotoxins and Their Relation to Other Water Quality Variables in Upper Klamath Lake, Oregon, 2007–09

By Sara L. Caldwell Eldridge, Tamara M. Wood, and Kathy R. Echols

Abstract

Phytoplankton blooms dominated by cyanobacteria that occur annually in hypereutrophic Upper Klamath Lake, Oregon, produce microcystins at concentrations that may contribute to the decline in populations of endangered Lost River (*Deltistes luxatus*) and shortnose (*Chasmistes brevirostris*) suckers. During 2007–09, water samples were collected from Upper Klamath Lake to determine the presence and concentrations of microcystins and cylindrospermopsins and to relate the spatial and temporal occurrences of microcystins to water quality and other environmental variables. Samples were analyzed for intracellular (particulate) and extracellular (dissolved) microcystins and cylindrospermopsins using enzyme-linked immunosorbent assays (ELISA). Samples contained the highest and most variable concentrations of microcystins in 2009, the year in which an earlier and heavier *Aphanizomenon flos-aquae*-dominated phytoplankton bloom occurred. Concentrations were lowest in 2008 when the bloom was lighter, overall, and delayed by nearly 1 month. Microcystins occurred primarily in dissolved and large (> 63 μm) particulate forms in all years of the study, and overall, concentrations were highest at MDT (the deepest site in the study) and HDB, although HDB was sampled only in 2007 and MDT was not sampled in 2008. Comparisons among daily median total microcystin concentrations; chlorophyll *a* concentrations; total, dissolved, and particulate nutrient concentrations; and nutrient ratios measured in 2009 and between 2007 and 2009 indicate that microcystin concentrations generally increase following the decline of the first *A. flos-aquae*-dominated bloom of each season in response to an increase in bioavailable nitrogen and phosphorus. Nitrogen fixation by *A. flos-aquae* early in the sample season appears to provide new nitrogen for growth of toxigenic *Microcystis aeruginosa*, whereas, later in the season, these species appear to co-exist. Understanding the ecological interactions between these species may be important for predicting periods of elevated cyanotoxin concentrations and has important implications for management of this lake.

Background

Two historically abundant, endemic fishes inhabiting the Upper Klamath Basin, Oregon, the Lost River sucker (*Deltistes luxatus*) and the shortnose sucker (*Chasmistes brevirostris*), were listed as endangered under the Federal Endangered Species Act in 1988 by the U.S. Fish and Wildlife Service following sharp declines in abundance, range reductions, and evidence that recruitment into the spawning population had decreased from historical levels (U.S. Fish and Wildlife Service, 1993; National Research Council, 2004). In addition to overharvest, habitat alteration, and the presence of nonnative fish species (U.S. Fish and Wildlife Service, 2002), the observed decline in these populations has been attributed to water quality degradation resulting from massive cyanobacterial blooms (Williams, 1988; Buettner and Scoppettone, 1990; Scoppettone and Vinyard, 1991; Perkins and others, 2000), which occur annually from June through October in Upper Klamath Lake. High rates of photosynthesis during bloom periods elevate lake water pH (9.5 and higher; Kann and Smith, 1999), and decomposition during bloom declines increases concentrations of un-ionized ammonia (> 0.5 mg/L) and dissolved nutrients (Hoilman and others, 2008; Lindenberg and others, 2009). This seasonal cycle of cyanobacterial growth and decline also causes oxygen concentrations to fluctuate from supersaturation to near anoxia (dissolved oxygen concentrations less than 1 mg/L; Wood and others, 2006), depending on the magnitude of the bloom decline (Kann and Welch, 2005).

In addition to contributing to poor water quality conditions through photosynthetic activity and bloom decomposition, several genera of bloom-forming, freshwater cyanobacteria also produce secondary metabolites that are toxic to a wide range of aquatic organisms, including fish (reviewed in Falconer, 1999). Hepatotoxic microcystins, first isolated from *Microcystis aeruginosa* (Krishnamurthy and others, 1986), are the most abundant and frequently occurring of these compounds (Carmichael and others, 1986; Yu, 1989; World Health Organization, 2006; Erdner and others, 2008) and have been implicated in human, livestock, domestic animal, and wildlife illness and death in more than 20 countries worldwide and in at least 36 U.S. states, including Oregon (Carmichael, 1994; Sivonen and Jones, 1999; Graham and others, 2009). More than 80 microcystin variants have been identified (Welker and Von Döhren, 2006), which vary in toxicity by an order of magnitude (Sivonen and Jones, 1999). Microcystins are nonribosomally synthesized cyclic heptapeptides produced by strains (subspecies) of many cyanobacterial genera, including *Microcystis* (Krishnamurthy and others, 1986), *Anabaena* (Harada and others, 1991), *Oscillatoria* (Meriluoto and others, 1989), *Gloeotrichia* (Carey and others, 2007), and *Pseudanabaena* (Oudra and others, 2002), all of which have been identified in water samples from Upper Klamath and Agency Lakes (Kann, 1997; Kann and Asarian, Aquatic Ecosystems Sciences LLC, unpub. data, 2008 and 2009; B.H. Rosen, U.S. Geological Survey, unpub. data, 2009; Eldridge and others, unpub. data, 2010). Blooms containing these genera may contain both toxic (with microcystin synthetase, *mcy*, gene clusters) and nontoxic (without the *mcy* genes) strains, which can not be distinguished from each other by microscopy. The nitrogen (N_2)-fixing (diazotrophic, ability to convert nitrogen gas to ammonia) *Aphanizomenon flos-aquae* generally comprises more than 90 percent of the cyanobacteria biovolume (Kann, 1997) in Upper Klamath Lake during periods of high bloom density. Members of the *Aphanizomenon* genus have been shown to produce cylindrospermopsins and several neurotoxins in laboratory cultures (Carmichael, 1997; Preussel and others, 2006; Graham and others, 2008), but they have not been shown to produce microcystins, and there is currently no evidence that *A. flos-aquae* produces toxins in the Upper Klamath or Agency Lakes (Carmichael, 2000). Therefore, although *A. flos-aquae* is the dominant cyanobacterium in this area, it is not likely to be the microcystin producer here.

The colony-forming, unicellular *M. aeruginosa* has been directly linked to microcystin occurrence in Upper Klamath Lake and in downstream reservoirs and was first reported in water samples from Upper Klamath Lake in July 1996 (Jacoby and Kann, 2007). The Oregon Department of Health and Health Canada used high performance liquid chromatography (HPLC) and enzyme-linked immunosorbent assays (ELISA) to show the presence of microcystins in dietary supplements produced from *A. flos-aquae* collected exclusively from Upper Klamath Lake (Gilroy and others, 2000; Lawrence and others, 2001). In a later study, Saker and others (2007) used multiplex polymerase chain reaction (PCR) to simultaneously detect a fragment of the microcystin-synthetase gene cluster (*mcyA*) and the partial 16S rRNA gene sequence specific to *Microcystis* in these supplements. The results of this study showed that *M. aeruginosa* was the most abundant microcystin-producer in the dietary supplements tested (*Planktothrix* sp. also was identified as a microcystin-producer) and supported previous reports that *Microcystis*

(mostly *M. aeruginosa*) co-occurs with *A. flos-aquae* in Upper Klamath Lake (Carmichael, 2000). *Microcystis* is not capable of nitrogen fixation, unlike *A. flos-aquae*, and is dependent on ammonia and other nitrogen sources for growth. Therefore, the availability of nitrogen may contribute to the occurrence of *Microcystis* colonies and microcystins in Upper Klamath Lake.

In 2007, the U.S. Geological Survey continued monitoring water quality throughout Upper Klamath Lake (a project that began in 2002; Wood and others, 2006) and began a preliminary study to determine the seasonal and spatial occurrence of cyanotoxins in the lake and to determine statistical relations between microcystin concentrations and other water quality variables. In addition, the U.S. Geological Survey conducted parallel studies to determine the pathological effects of cyanotoxins on juvenile Lost River and shortnose suckers in Upper Klamath Lake (VanderKooi and others, 2010). Between July and September 2007, histopathology consistent with microcystin exposure (Malbrouck and Kestemont, 2006) was identified in 49 percent (n = 47) of age-0 suckers collected from 11 shoreline locations throughout the lake (VanderKooi and others, 2010). Multiple organ necrosis was identified in a greater percentage of individuals captured in the northern region of the lake than in the southern region. In summer 2008 (July–September), age-0 suckers (n = 103) collected from five geographic areas in the lake were examined for histopathology. Evidence for organ damage was absent in two areas and observed in less than 20 percent of fish collected in other areas. As in the previous year, fish captured in the northern region of the lake exhibited the highest occurrence of organ damage. These regional differences may be due to variation in the dose, exposure duration, and time since exposure to toxins in the lake. A gut analysis of juvenile fish collected in 2008 (n = 45) showed that all suckers observed had ingested chironomid (midge) larvae which, in turn, appeared to contain colonies of *M. aeruginosa* and filaments of *A. flos-aquae* in their digestive tracts (*M. aeruginosa* was found in all ingested chironomid larvae and *A. flos-aquae* was found in approximately 20 percent of the larvae; B.H. Rosen, U.S. Geological Survey, unpub. data, 2008). The juveniles examined had completed the ontogenetic (developmental) shift to benthic feeding, so these chironomid larvae were most likely ingested from lake sediments. Histological examinations of these fish revealed numerous gastro-intestinal lesions consistent with microcystin exposure (Malbrouck and Kestemont, 2006), which were observed regardless of whether liver necrosis also was present (VanderKooi and others, 2010). The observed histopathology did not indicate a bacterial or parasitic etiology, although infection from an undetected virus or chronic effects of ammonia toxicity could not be ruled out. These fish most likely were exposed to microcystins by ingestion and not by absorption of these compounds through their gills.

Cyanotoxin concentrations may vary widely between sites and depths within a single lake or during a sampling season and between years. Such spatiotemporal variation generally results from indirect environmental influences on the presence and abundance of toxigenic strains (Kurmayer and others, 2003; Kurmayer and Christiansen, 2009), and from the direct effects of the environment on cellular rates of microcystin production, which are determined by cyanobacterial community structure, growth stage, and nutrient dynamics (Reynolds, 1998; Jacoby and others, 2000). Changes in cellular microcystin production rates and the abundance of toxigenic *M. aeruginosa* strains have been documented on a seasonal basis in reservoirs downstream of Upper Klamath Lake (Kann and Corum, 2009, 2010; Bozarth and others, 2010). Determining which environmental factors, direct or indirect, have the greatest effect on naturally occurring communities of mixed cyanobacteria populations is difficult. However, studies have shown that microcystin production is favored by factors that regulate population growth (Orr and Jones, 1998), including limitation by phosphorus (Trimbee and Prepas, 1987; Jacoby and others, 2000), nitrogen (Jones and Jones, 2002; Downing and others, 2005; Moisander and others, 2009), and light (Wicks and Thiel, 1990; Wiedner and others, 2003), among others. Although microcystins are contained primarily within the cells that produce them, dissolved microcystins can be detected during early stages of bloom development (Sedmak and Elersek, 2006) because cellular excretion and lysis occur continually throughout the growth cycle (Hughes and others, 1958). However, the highest concentrations of dissolved microcystins commonly are detected during bloom senescence and decomposition (Park and others, 1998). Therefore, in the present study, an elevated concentration of dissolved microcystins was considered indicative of a decline in the *M. aeruginosa* population.

Study Area

Upper Klamath Lake, located within the Klamath Graben structural valley at the base of the Cascade Mountains (eastern slope) in south-central Oregon, is a naturally occurring, large and shallow water body with a surface area of 232 km² and an average depth of 2.8 m (fig. 1). More than 90 percent of the lake is less than 4 m deep, but the western shoreline between Eagle Ridge and Buck Island (crossing the entrance to Howard Bay) reaches depths of up to 15 m. The Williamson River, which enters the lake from the north, contributes, on average, approximately 46 percent of the total inflowing water to the lake annually (Johnson, 1985). The lake's drainage basin is 9,415 km² and is composed of phosphorus-rich, volcanic soils. The Upper Klamath Lake system has been eutrophic since at least the mid-1800s (the earliest known records), but major changes in land use and hydrology of the watershed and lake over the past century, including forest clear-cutting, cattle grazing in upstream flood plains, degradation of riparian corridors, and the conversion of neighboring wetlands to flood-irrigated pasture and agricultural fields (Klamath Tribes, 1994), have intensified eutrophication and increased the biomass of cyanobacteria (primarily of a single species, *A. flos-aquae*) that bloom during the summer and autumn (Bortelson and Fretwell, 1993; Bradbury and others, 2004; Eilers and others, 2004). Although the lake occurs naturally, it is artificially controlled. Upper Klamath Lake has been the primary water source for the Klamath Project, which has supplied water to agricultural areas within the Upper Klamath Basin since the completion of the Link River Dam at the southern outlet of the lake in 1921 (Bureau of Reclamation, 2000). This has allowed regulation of lake water levels and volume that has resulted in more extreme fluctuations in annual surface water elevation (as much as 1 m between early spring and late summer), although, in recent years, the overall maximum and minimum elevations have been similar to pre-dam conditions.

Purpose and Scope

The purpose of this report is to characterize and compare cyanotoxin concentrations in water samples collected from Upper Klamath Lake during the 2007 through 2009 field seasons. The physical and chemical characteristics of study sites are summarized to determine if relations exist among the spatial and temporal distributions of microcystins and other environmental variables in Upper Klamath Lake, including dissolved and total nutrients; chlorophyll *a*; particulate carbon, nitrogen, and phosphorus; water temperature; pH; dissolved oxygen; wind speed; site depth; and water column stability (relative thermal resistance to mixing, RTRM). Nutrient, chlorophyll *a*, continuous monitor, and meteorological data obtained through the long-term monitoring program are included here to provide context for microcystin analyses, but are described in Kannarr and others (2010). Results of this study will contribute to understanding the environmental influences on the occurrence of cyanotoxins, specifically microcystins, in Upper Klamath Lake, which is critical for effective lake management and understanding the causes of apparent high mortality rates in juvenile Lost River and shortnose suckers.

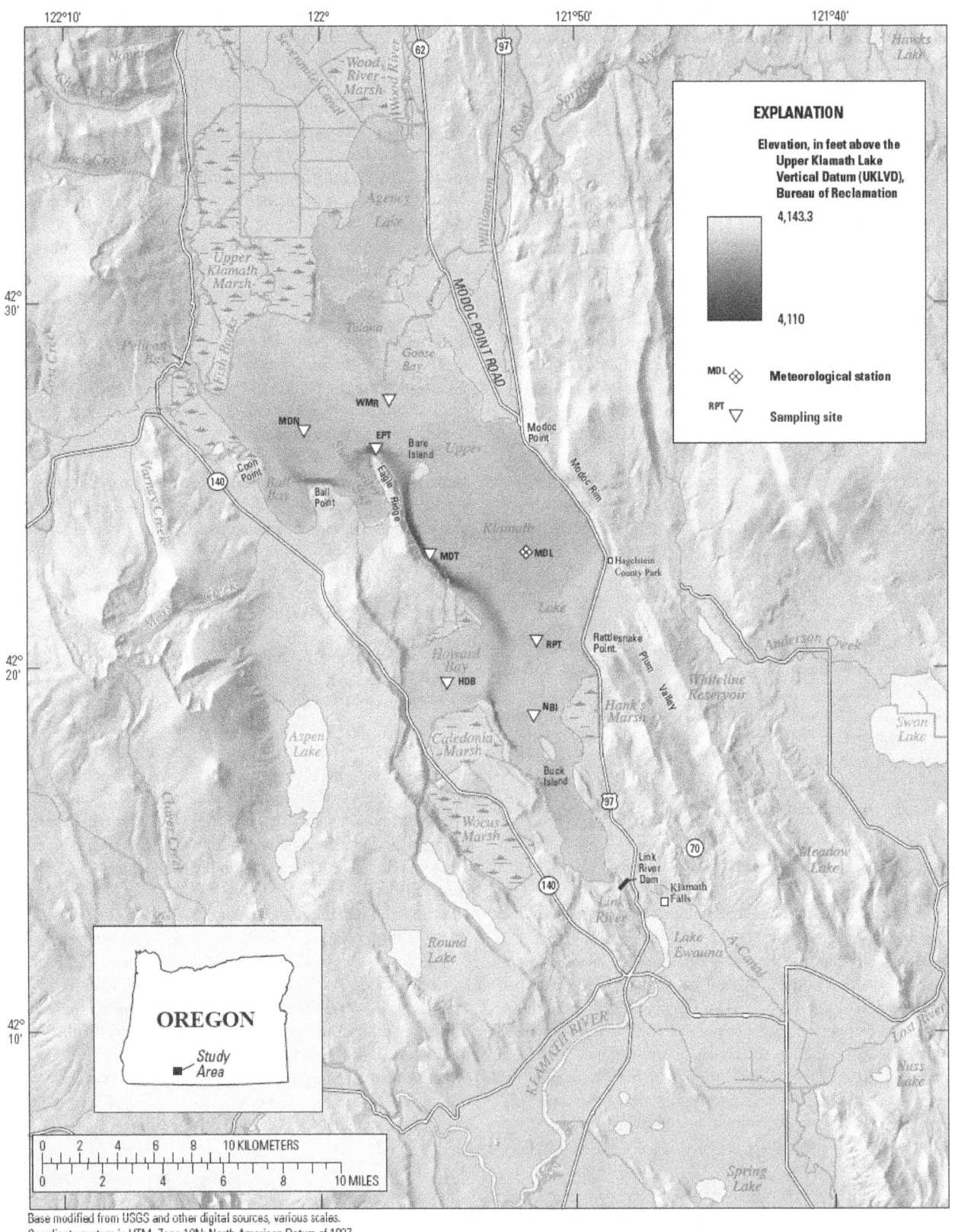

Figure 1.   Locations of meteorological station and cyanotoxin sampling sites, Upper Klamath Lake, Oregon, 2007–09.

Methods

Water Quality Data Collection

Water column samples were collected from different sites and at various sampling intervals between 2007 and 2009 to create overlap between water quality data collection and the U.S. Geological Survey's juvenile sucker sampling effort; resource constraints also have limited sampling in some years. In 2007, water samples were collected at six sites on Upper Klamath Lake that were co-located with continuous water-quality monitoring sites according to established collection and quality control protocols (U.S. Geological Survey, variously dated) for the analysis of cyanotoxins (microcystins and cylindrospermopsins), chlorophyll *a*, dissolved nutrients (orthophosphate [DIP], ammonia and nitrite plus nitrate [DIN]), total phosphorus (TP), and total nitrogen (TN). Samples for cyanotoxin analysis were collected monthly (July 9 and 10, July 31 and August 1, September 4 and 5, and October 16 and 17) at sites MDN, WMR, EPT, MDT, RPT, and HDB (table 1; fig. 1); nutrient and chlorophyll *a* samples were collected weekly from these sites from mid-May to mid-November. In 2008, the sampling frequency for cyanotoxin analysis was increased from monthly to biweekly and occurred between June 30 and September 22 at fewer sites—MDN, WMR, MDL, and NBI—than in the previous year. Also in 2008, samples for nutrient and chlorophyll *a* analyses were collected weekly from mid-May to early October, but at only two of the four cyanotoxin sample sites, MDN and WMR. In 2009, weekly sampling for nutrient and chlorophyll *a* analyses began in mid-May (as in the previous 2 years) and occurred concurrently with cyanotoxin sample collection (also weekly) between June 16 and September 14 (weekly sampling for nutrient and chlorophyll *a* analyses continued through September 28.) All samples were collected from five sites in 2009—MDN, WMR, EPT, MDT, and RPT (table 1; fig. 1). Additional samples were collected between June 16 and August 31, 2009, for analyses of total particulate carbon (TPC), total particulate phosphorus (TPP), total particulate nitrogen (TPN), and particulate inorganic phosphorus (PIP).

Water samples for chlorophyll *a*, TN, and TP analyses were integrated by collecting lake water in two 1-L vented bottles held in a weighted cage that was dropped at a constant rate from the surface to 0.5 m from the sediment at shallow sites (< 10.5 m), and to 10 m from the surface at deep sites (> 10.5 m). Samples were mixed in a churn splitter and divided into separate fractions for each type of analysis. Samples analyzed for dissolved nutrients (ammonia, orthophosphate, and nitrite plus nitrate) were collected at discrete depths to identify sources (benthic or water column) of these nutrients for the long-term monitoring project. To do this, a hose was lowered to one-half the water column depth at shallow sites and at two points, one-quarter and three-quarters the water column depth, at deep sites. Lake water was drawn into the hose by a peristaltic pump, passed through a 0.45-μm capsule filter attached to the hose end, and collected into sample bottles. Total particulate carbon and nutrient (TPC, TPN, and TPP) samples were also collected at discrete depths and in the same manner as the dissolved nutrient samples, but with the capsule filter removed. Water samples for particulate carbon and nutrient analyses were filtered at the U.S. Geological Survey Klamath Falls Field Station on 25-mm (precumbusted for TPC and TPN) or 47-mm (for TPP and PIP), 0.7-μm pore size, glass microfiber filters (GF/F, Whatman, Inc., Piscataway, N.J.) and shipped overnight to the University of Maryland Chesapeake Bay Laboratory (CBL), Solomons, Maryland. Once received, samples for TPC and TPN analyses were processed according to EPA Method 440.0 (U.S. Environmental Protection Agency, 1997), and TPP concentrations were analyzed according to Aspila and others (1976). Samples for the measurement of PIP were extracted in an acidic medium and analyzed according to the method of Aspila and others (1976). Further processing and analyses of chlorophyll *a*, total nutrients, and dissolved nutrients are described in Kannarr and others (2010). Nutrient (total and dissolved) and chlorophyll *a* samples collected in 2009 were processed as in 2008.

From June 16 to August 3, 2009, and during all samplings in 2007 and 2008, cyanotoxin samples were collected by depth integration (described above), transferred from the churn splitter to 1-L amber, high density polyethylene (HDPE) bottles, and immediately stored on ice. Samples collected on August 10, 2009, and thereafter were collected by depth integration at the shallower sites, MDN, WMR, and RPT, and at two points, one-quarter and three-quarters the water column depth, at the deeper sites, MDT and EPT, using a hose and peristaltic pump (the same method as for particulate nutrient sample collection). Samples collected in all years were shipped overnight to the U.S. Geological Survey, Columbia Environmental Research Center (CERC), Columbia, Mo. Once received, samples were kept on ice in a walk-in cooler and processed within 24 hours. Samples for quality-assurance determination also were collected, and the results of this analysis are presented in appendix A.

Table 1.   Summary of samples or data collected for analyses of cyanotoxin occurrence and other water quality parameters, Upper Klamath Lake, Oregon, 2007–09.

[Locations of sample sites are shown in figure 1. Samples for nutrient (total and dissolved) and chlorophyll *a* analyses and continuous monitoring data collected on the same days as cyanotoxin samples are included. Continuous monitoring data were collected hourly and reported as daily median values on toxin sample dates. Water samples collected in 2007 do not include chlorophyll *a*. Water samples were not analyzed for dissolved toxins in 2007. Italicized numbers are depth-integrated samples. All other samples or data were collected at discrete depths. Cyanotoxin samples were collected by depth integration (June 16–August 3) and as point samples (August 10–September 14) in 2009 only. **Symbol:** –, no data]

Sample site	Median depth (meters)	2007					2008				
		Cell associated toxins	Dissolved toxins	Total nutrients	Dissolved nutrients	Continuous monitors	Cell associated toxins	Dissolved toxins	Total nutrients chlorophyll *a*	Dissolved nutrients	Continuous monitors
MDN	3.4	*4*	–	*4*	4	–	7	7	7	7	–
MDN Upper	0.85	–	–	–	–	4	–	–	–	–	7
MDN Lower	2.5	–	–	–	–	4	–	–	–	–	7
WMR	1.3	*4*	–	*4*	4	4	7	7	7	7	7
EPT	11.1	*4*	–	*4*	–	–	–	–	7	–	–
EPT Upper	2.7	–	–	–	4	4	–	–	–	7	7
EPT Lower	8.1	–	–	–	4	4	–	–	–	7	7
MDT	13.6	*4*	–	*4*	–	–	–	–	7	–	–
MDT Upper	3.2	–	–	–	4	4	–	–	–	7	7
MDT Lower	9.7	–	–	–	4	4	–	–	–	7	7
MDL	3.5	–	–	–	–	–	7	7	–	–	–
RPT	2.5	*4*	–	*4*	4	4	–	–	–	–	7
NBI	2.0	–	–	–	–	4	7	7	–	–	7
HDB	1.3	*4*	–	*4*	4	4	–	–	–	–	7

Sample site	Median depth (meters)	2009					
		Cell associated toxins	Dissolved toxins	Total nutrients chlorophyll *a*	Dissolved nutrients	Particulate, carbon/ nitrogen/ phosphorus	Continuous monitors
MDN	3.4	*14*	*14*	*14*	14	12	–
MDN Upper	0.85	–	–	–	–	–	14
MDN Lower	2.5	–	–	–	–	–	14
WMR	1.3	*14*	*14*	*14*	14	12	14
EPT	11.1	*8*	*8*	*14*	–	–	–
EPT Upper	2.7	*6*	*6*	–	14	12	–
EPT Lower	8.1	*6*	*6*	–	14	12	14
MDT	13.6	*8*	*8*	*14*	–	–	–
MDT Upper	3.2	*6*	*6*	–	14	12	14
MDT Lower	9.7	*6*	*6*	–	14	12	14
MDL	3.5	–	–	–	–	–	–
RPT	2.5	*14*	*14*	*14*	14	12	14
NBI	2.0	–	–	–	–	–	14
HDB	1.3	–	–	–	–	–	–

Continuous water-quality monitors (YSI, Inc., Yellow Springs, Ohio) were deployed each year at sites shown in table 1 to measure water temperature, dissolved oxygen, pH, and specific conductance according to Lindenberg and others (2009). Monitors were positioned vertically 1 m from the lake bottom at all sites, except site WMR, and an additional sonde was placed on the same mooring 1 m from the lake surface at the deeper sites, MDN, EPT, and MDT. Water column stability was calculated as the median relative thermal resistance to mixing (RTRM) using data collected by the upper and lower sonde at site MDN. RTRM was determined by comparing the water column density gradient (based on the temperature difference between the upper and lower sondes) to the density difference between 4 and 5°C (Jones and Welch, 1990; Kann and Welch, 2005). In 2009, only the lower sonde (1 m from the bottom) was deployed at site EPT. The depth at site WMR is less than 2 m, so the sonde at this site was placed horizontally at one-half the water column depth. Calibration, maintenance, and data handling were performed following standard procedures outlined in Wagner and others (2006). Meteorological data were collected from the floating station at site MDL as described in Lindenberg and others (2009) and Kannarr and others (2010).

Cyanotoxin Analysis

Water samples were fractionated (filtered) at CERC to determine the relative contributions of different phytoplankton size classes to the total cyanotoxin concentration in each sample. Samples were first filtered with a 63 µm sieve to isolate large cyanobacterial filaments and colonies. The small fraction, 1.5–63 µm, representing smaller forms, was collected by filtering the 63-µm fraction filtrate onto pre-weighed ProWeigh™ glass fiber filters (Environmental Express, Mt. Pleasant, S.C.). The filtrate from this step was considered the dissolved fraction and retained for determination of dissolved microcystin (< 1.5 µm) concentrations. Prior to extraction, both the large and small particulate fractions were freeze-dried and weighed. In 2007, samples were extracted three times with 5 mL of 50 percent aqueous methanol and 0.1 percent trifluoroacetic acid for 5 minutes by ultrasonication. The extracts were centrifuged at 10,000 revolutions per minute (rpm) for 15 minutes, and the supernatant was filtered through a 0.45-µm nylon syringe filter or a 0.45-µm UniPrep™ syringeless glass microfiber (GMF) filter (Whatman Inc., Piscataway, N.J.). In 2008 and 2009, freeze-dried biomass (containing the large particulate fraction) and filters (containing the small particulate fraction) were extracted using a Dionex Accelerated Solvent Extraction System, ASE 200 (Dionex Corporation, Sunnyvale, Calif.) in 5-mL sample cells and using the parameters reported in Aranda-Rodriguez and others (2005). The cells were prepared by tamping a glass fiber filter (Fisher G2; 1 µm, cut to size with a cork borer) into the exit end of each cell and filling it half full of glass beads (Kimble Kimax KG-33; 3 mm) for filter extraction or with Hydromatrix for extraction of raw water filtrate.

The samples were dried under nitrogen to remove methanol and resuspended in 10 mL deionized water. Diluted (1/100) extracts were analyzed in 96-well microtiter plates for determination of microcystin and cylindrospermopsin concentrations using congener-independent enzyme-linked immunosorbent assays (ELISA; kit 520011, Abraxis, LLC, Warminster, Pa.) following the manufacturer's protocol. Absorbances of the samples at 450 nm were determined within 15 minutes after addition of the final solution. Concentrations were determined from a regression of the mean absorbance of calibration standards, and were analyzed in duplicate (duplicate blanks were included) at the same time as the samples. Total microcystin or cylindrospermopsin concentrations were determined by summing the particulate and dissolved concentrations (Graham and Jones, 2007). Concentrations in all size fractions were calculated volumetrically to facilitate comparisons and to calculate total microcystin concentrations. However, because the route of exposure to affected suckers in Upper Klamath Lake is likely through ingestion, concentrations of cell associated microcystins also were expressed as mass per dry weight of suspended solids. Dissolved fractions of both toxins were not analyzed in 2007, and cylindrospermopsin concentrations were not measured in 2009. Therefore, total concentrations described for 2007 may be underestimates. Samples collected in 2007 and 2008 contained cylindrospermopsins near or less than the detection limit and are not described further in this report. The detection limit for the microcystin assay is 0.10 ppb (µg/L), and the detection limit for the cylindrospermopsin assay is 0.05 ppb. Values less than 0.1 µg/L that are not censored (as < MQL) are considered detections, although they appear to be less than the detection limit since the aqueous concentrations were calculated from the total extracted biomass and sample volumes. Cyanotoxin data are available in appendix B.

Data Analysis

Maximum-likelihood estimation (MLE) of summary statistics for datasets with censored values have been shown to produce estimates with large bias and poor precision for small (n < 15) sample sizes (Gleit, 1985). Therefore, for this reason and for simplicity, values equal to or less than the detection limit were considered equal to the detection limit for determination of median values. This method also was used for statistical analysis, because only 13 percent of the samples (all years combined) collected between July 7 and September 14 (the sampling period common to all study years, hereafter referred to as July–September) contained dissolved microcystins less than the detection limit, and no samples contained microcystins in the large particulate fraction at concentrations less than the detection limit during that time; small particulate (cell associated) microcystin data were not used in statistical analysis.

Changes in cyanotoxin, nutrient, and chlorophyll *a* concentrations over time were compared by calculating the daily median values of these variables measured in water samples from all sites on each sample date. Analyses of water temperature, pH, and dissolved oxygen concentration or percent saturation were based on the median value of 24 hourly measurements made at all sites on each sample date. Interannual comparisons of cyanotoxin data were based on daily median values of concentrations determined in samples from all sites on each sample date between July and September. Inter- and intra-annual comparisons of cyanotoxin concentrations measured between sites were based on daily median values at each site between July and September.

Correlations between microcystin concentrations in the dissolved and large particulate (> 63 μm) fractions and environmental variables were determined using Spearman rank order correlation analysis; correlations were considered significant at *p* < 0.05 and near significance at *p* < 0.1. Due to the low concentration (maximum values between 0.03 and 0.07 μg/L), relative to those in the dissolved and large particulate fractions, microcystins within the small particulate (1.5–63 μm) fraction and cylindrospermopsins were not included in comparisons with environmental variables. All microcystin, chlorophyll *a*, and nutrient data collected at sites MDN and WMR, the only sites commonly sampled from 2007 to 2009, were used in interannual correlations. Interannual correlations with data collected from continuous monitors (water temperature, pH, and dissolved oxygen concentration or percent saturation) were based on the median of 24 hourly measurements collected on each sample date. For interannual correlations, all data collected in 2007 and 2008 were included, but only data collected on corresponding sample dates in 2009 were used in order to weight the years evenly. Intra-annual correlations were based on median values of data collected at all sites on each sample date in 2009. Correlations with wind speed were based on the median of 24 hourly measurements made on each sample date at meteorological station MDL.

Interannual Variability, 2007–09

Seasonal and interannual fluctuations in Upper Klamath Lake water quality and associated environmental parameters frequently reflect changes in *A. flos-aquae*-dominated phytoplankton biomass as measured by the photosynthetic pigment, chlorophyll *a*, and may be important for understanding the occurrence of cyanotoxins in the lake. Indicators of bloom decline and cell senescence include decreases in chlorophyll *a* concentrations, pH, and dissolved oxygen concentrations that accompany decreased photosynthetic activity (and active decomposition), and an increase in dissolved nutrient concentrations as cells lyse in the water column. In addition, year-to-year variability in the

severity of these changes may be strongly related to weather (Perkins and others, 2000; Kann and Welch, 2005). Periods of bloom decline often coincide with periods of sustained high temperature and low wind, which increases water column stability (Kann and Welch, 2005). In years when a severe mid-season bloom decline is observed, the decline is accompanied by the seasonal maximum water temperature (Wood and others, 2006; Hoilman and others, 2008; Lindenberg and others, 2009; Kannarr and others, 2010).

Chlorophyll *a* and Nutrients

Water quality and meteorological data collected in 2007 and 2008 and used in this study are discussed in Kannarr and others (2010). Temporal changes in the median values of a subset of water quality parameters that are hypothesized to influence the presence of, or short-term changes in the occurrence of, cyanotoxins are summarized in figure 2 of the current report. Comparison of median values show that chlorophyll *a* concentrations were generally higher from July to September in 2008 than at the same sites sampled in 2009. Although the difference in seasonal median values was not large, 2008 samples collected between June 16 and September 15 exhibited a wider chlorophyll *a* concentration range (median values from 206 to 27 μg/L in 2008 and from 164 to 3.4 μg/L in 2009), but did not decrease sharply mid-season, as in 2009, signifying a major bloom decline. Chlorophyll *a* data analyzed in 2007 were not reported because the quality of the data was poor (Kannarr and others, 2010). Median values of total nitrogen and phosphorus were highest in 2007 and comparable between 2008 and 2009. The annual median ammonia concentration measured in 2009 was more than double that of 2007 and more than 10 times the median value for 2008. This large difference in overall ammonia concentrations between years, in contrast to the smaller difference in orthophosphate, accounts for the higher seasonal median ratio of dissolved inorganic nitrogen to dissolved inorganic phosphorus (DIN:DIP) observed in 2009. Between July and September, the 2009 median DIN:DIP (3.30) was nearly three times the median value observed in 2007 (1.19) and six times the 2008 median value (0.55).

Total nitrogen to total phosphorus (TN:TP) ratios indicate potential for phytoplankton nutrient limitation, and low TN:TP values (less than 29 by weight; Smith, 1983) are often associated with cyanobacterial success (Smith, 1983; Paerl, 1988; Jacoby and others, 2000). TN:TP ratios greater than 17, by weight, and chlorophyll *a* to total phosphorus ratios (chlorophyll *a*:TP) greater than 1, by weight, have been measured in phosphorus-limited lake systems (Forsberg and Ryding, 1980; White, 1989; Graham and others, 2004), including Upper Klamath Lake, where previous data generally show potentially phosphorus-limiting conditions early in the sampling season (late June and early July; Lindenberg and others 2009). In a study of 30 waste-receiving lakes sampled

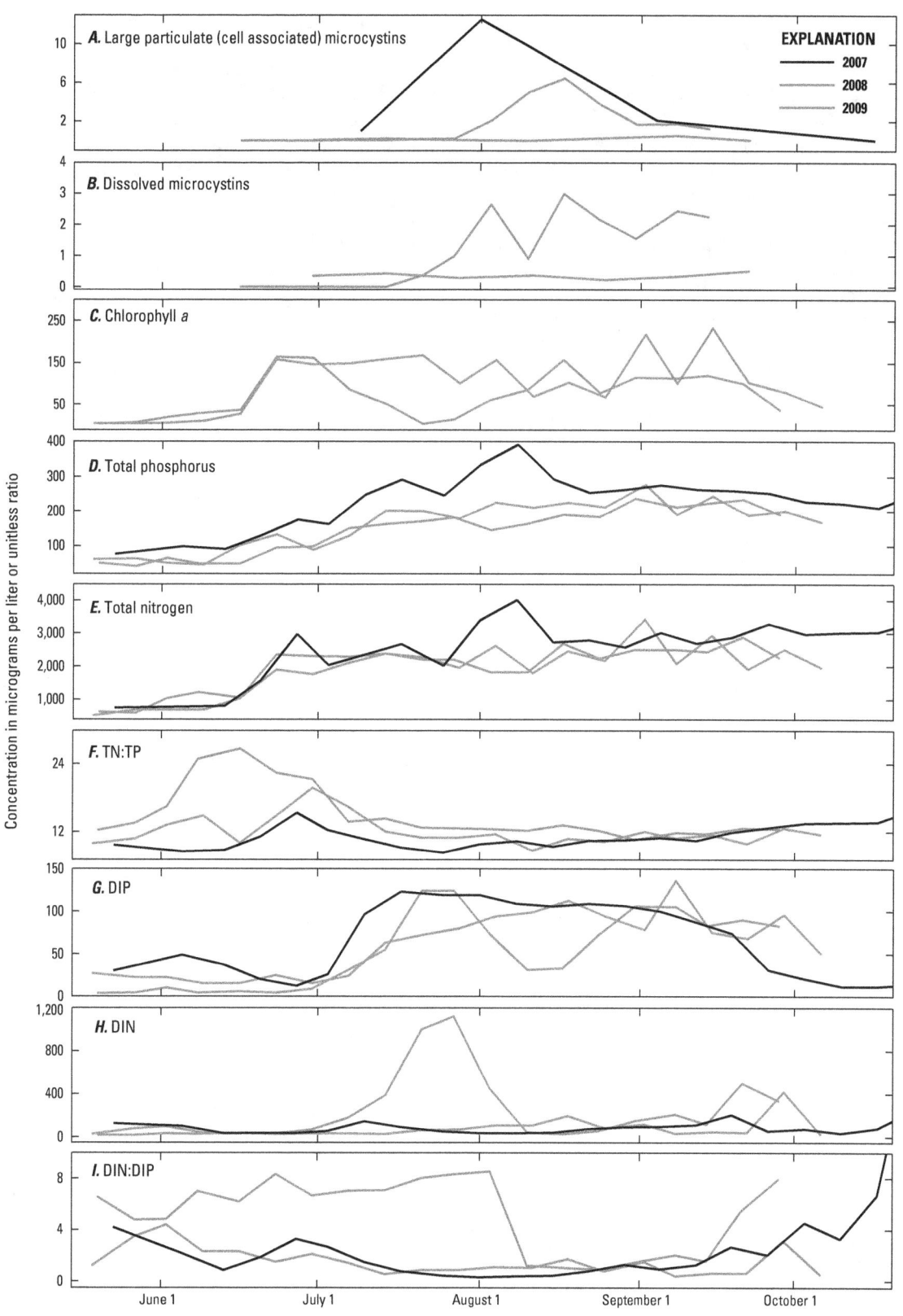

Figure 2.   Median concentrations of large (> 63 µm) particulate (cell associated) microcystins, dissolved microcystins, chlorophyll *a*, total nutrients (nitrogen and phosphorus), the ratio of total nitrogen to total phosphorus, dissolved nutrients, and the ratio of dissolved inorganic nitrogen to dissolved inorganic phosphorus in water samples from Upper Klamath Lake, Oregon, 2007–09. Source for 2007–08 data: Kannarr and others, 2010.

with high frequency over a 6-year period (4,500 samples), nitrogen was the most growth-limiting nutrient at TN:TP ratios less than 10 (Forsberg and Ryding, 1980). In addition, laboratory experiments have shown decreased amounts of microcystins under low phosphorus conditions (Sivonen and Jones, 1999), and recent studies using environmental samples show strong correlations between microcystin concentrations or *M. aeruginosa* cell densities and levels of total phosphorus (microcystin-LR in particulate form; Kotak and others, 2000; Downing and others, 2001; Chen and others, 2009) or total nitrogen (Downing and others, 2001). Seasonal median values of TN:TP ratios in Upper Klamath Lake between July and September were near 10 in all years (fig. 2), and chlorophyll *a*:TP ratios during this time do not support phosphorus (P)-limitation, because most values were less than 1 (fig. 3). However, consistent with results from previous years (Lindenberg and others, 2009), ratios of TN:TP were higher between mid- to late-June in all years (fig. 2), suggesting that, if P-limitation does occur, it is most likely during the first bloom of predominantly *A. flos-aquae*.

Dissolved Oxygen, pH, and Temperature

Dissolved oxygen and pH data collected from continuous monitors provide high-resolution measurements of water quality resulting from growth and decline of the *A. flos-aquae*-dominated bloom. Rapid growth phases were indicated in 2007 by supersaturated dissolved oxygen concentrations prior to the first week in July, between the last week in July and first week in August, and after the first week in September. Undersaturated dissolved oxygen concentrations (less than about 90 percent) and slightly lower pH signified bloom declines for 2 weeks in mid-July, at the end of August, and into early September (fig. 4; Kannarr and others, 2010). In 2008, undersaturated dissolved oxygen concentrations and lower pH in late May indicated a minor bloom decline, although the first major decline did not occur until mid-August and for a shorter time period in mid-September (fig. 4; Kannarr and others, 2010). Based on the degree of undersaturation and the relatively small decrease in pH, bloom declines in 2007 and 2008 were less severe than in 2009, when a large,

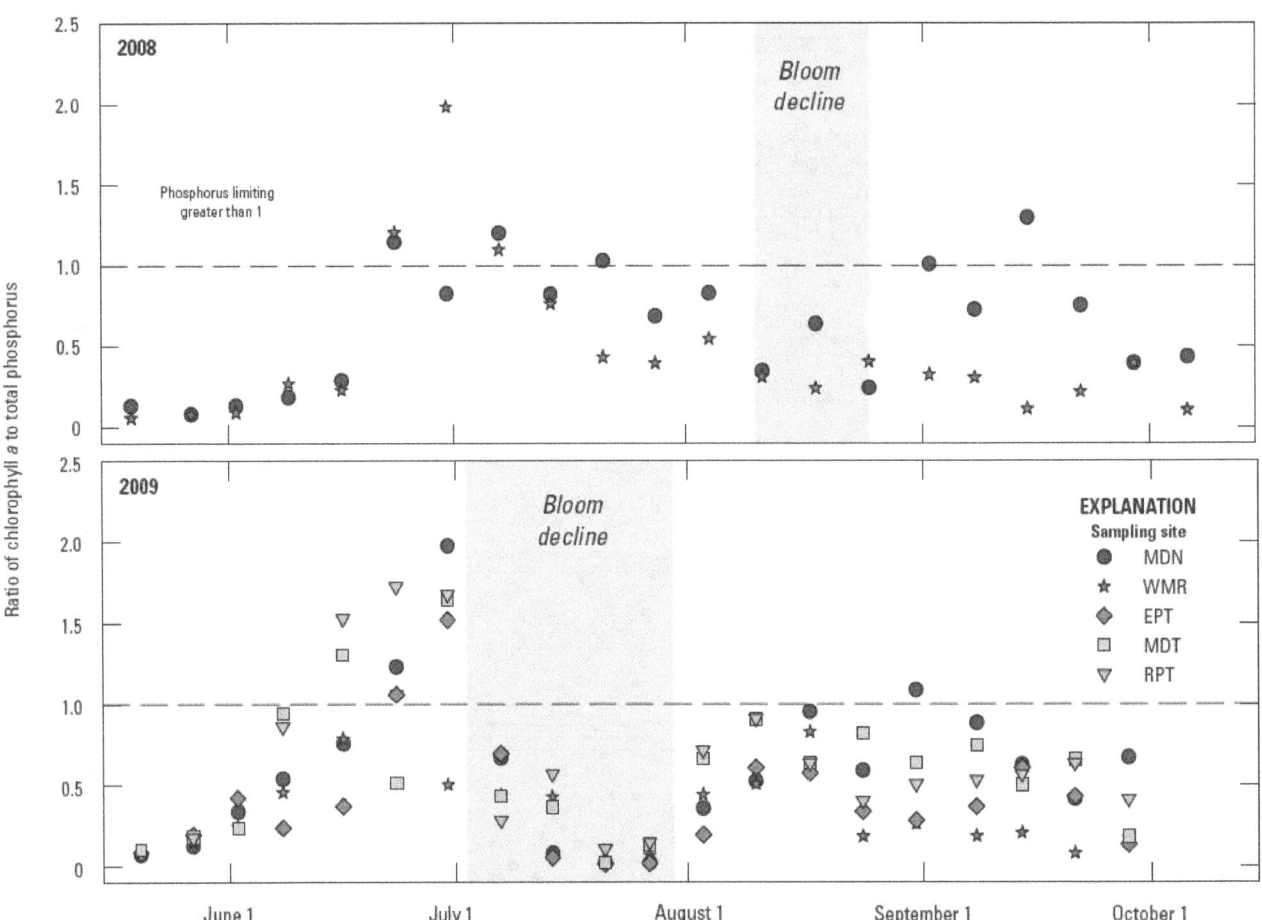

Figure 3.   Ratio of chlorophyll *a* to total phosphorus concentrations in water samples collected from Upper Klamath Lake, Oregon, 2008–09. Bloom decline indicates the period of minimum chlorophyll *a* concentrations.

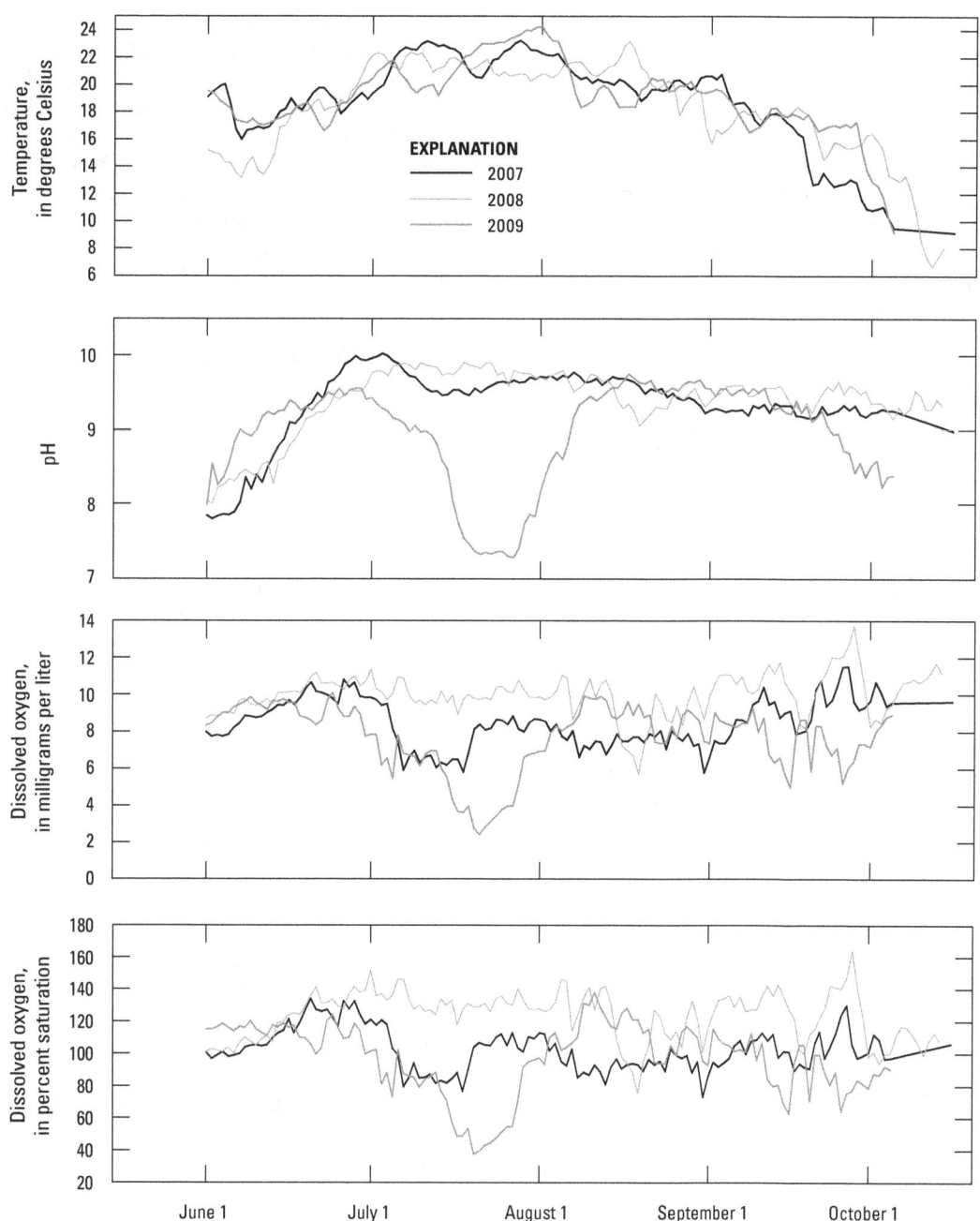

Figure 4. Median water temperature, pH, dissolved oxygen concentrations, and dissolved oxygen percent saturation at sites in Upper Klamath Lake, Oregon, 2007–09. Source for 2007–08 data: Kannarr and others, 2010.

mid-season decline occurred during the last 2 weeks in July (fig. 4), coinciding with the seasonal minimum in dissolved oxygen (2.45 mg/L) and pH (7.30; fig. 4). As previously noted, when a large mid-season bloom decline occurs, minimum dissolved oxygen concentrations and pH values are observed within a few days of the seasonal maximum water temperature (Hoilman and others, 2008; Lindenberg and others, 2009; Kannarr and others, 2010). Therefore, the bloom cycle in each year of this study can be qualitatively summarized between

July and September as: two declines in 2007 that culminated in mid-July and early September, a relatively extended, mild bloom decline in 2008, which occurred during the last 3 weeks of August, and a major bloom decline in 2009 that progressed throughout July and was most severe in the second half of that month. This description is broadly consistent with results of chlorophyll *a* analysis in samples collected once every 2 weeks from a larger set of eight sites in Upper Klamath Lake by the Klamath Tribes, Chiloquin, Oregon (Kann, 2010).

Cell Associated Microcystins

Preliminary data suggest that exposure of juvenile suckers to microcystins in Upper Klamath Lake is through ingestion, either through the food chain or by direct consumption of toxigenic cyanobacteria strains (VanderKooi and others, 2010). Therefore, it may be important to understand how microcystin concentrations vary between the particulate (intracellular, or cell associated) and dissolved (extracellular) phases. In all years, microcystins in the large (> 63 µm) particulate fraction, representing filamentous and (or) large planktonic colonies, were more concentrated than microcystins within the small (1.5–63 µm) particulate fraction (table 2). Between July and September, large particulate microcystin concentrations, expressed volumetrically, were highest in 2007 (median = 1.87 µg/L) and were more variable with a higher maximum in 2009 (range between 0.001 and 24.4 µg/L; table 2). Large particulate microcystin concentrations were lowest in 2008, ranging between 0.01 and 1.55 µg/L (median = 0.20 µg/L). In 2009, microcystins were dominantly in the large particulate fraction on most sample dates at all sites (fig. 5), indicating the presence of a microcystin-producing bloom that year. Interestingly, 2008 samples contained the lowest microcystin concentrations in the study, but the dissolved fraction comprised a larger percentage of total microcystins that year. This may be due to the presence of fewer *M. aeruginosa* colonies that year (J. Kann, Aquatic Ecosystem Sciences, LLC, unpub. data, 2009). However, given the relatively low concentrations of dissolved and particulate microcystins in 2008, this dominance of dissolved microcystin concentrations may not be significant.

Results of monthly sampling in 2007 provided only a general temporal pattern of particulate microcystin concentrations that year, but peak concentrations (> 12 µg/L) were observed at four of the six sites sampled on July 31 and August 1, at least 10 days earlier than when particulate fraction microcystin concentrations at all sites peaked in 2009 on various dates between August 10 and 31 (fig. 2). Two sites, MDN and WMR (fig. 1), were sampled for cyanotoxins in all 3 years, so the data collected from these sites can be compared explicitly. Samples from site MDN contained maximum large particulate microcystin concentrations of 17.4, 0.33, and 11.8 µg/L, measured on August 1, 2007, September 8, 2008, and August 17, 2009, respectively. Maximum concentrations measured in samples from site WMR were 7.35, 1.13, and 6.23 µg/L and occurred on September 5, 2007, September 8, 2008, and August 10, 2009, respectively (table 3). It also is noteworthy that on the last sample dates in 2007 (October 17), 2008 (September 22), and 2009 (September 14), concentrations were all less than 1 µg/L, less than or equal to 1.44 µg/L, and less than or equal to 2.26 µg/L, respectively.

This indicates a significant reduction in microcystin-bearing cells in the water column by the end of September (fig. 2).

Expressed as mass per dry weight of suspended solids, the median concentration of microcystins in the large particulate fraction between July and September was highest in 2009 (0.15 µg/mg), lowest in 2008 (0.02 µg/mg), and between these values in 2007 (0.09 µg/mg; table 2). This suggests that the contribution of toxigenic *M. aeruginosa* to the phytoplankton community was greatest in 2009 and least in 2008.

Dissolved Microcystins

Samples collected in 2009 contained higher (by more than five times) and more variable concentrations of dissolved microcystins (median = 1.49 µg/L) than the 2008 samples (median = 0.27 µg/L; table 2), although 15 percent of the samples collected in 2009 contained dissolved microcystins less than the detection limit (dissolved microcystins were detected in all 2008 samples, but sampling began later that year). In 2008, microcystins were primarily in the dissolved fraction (85.8 percent), but in 2009, dissolved microcystins comprised less than one-half of the total microcystin concentration (43 percent of the total concentration was in the dissolved form, and 56 percent was in the large particulate fraction; table 2; fig. 5). Concentrations in 2008 remained less than 1 µg/L at all sites except MDN, where the concentration was just over 1 µg/L on 2 sample dates, July 14 (1.10 µg/L) and September 22 (1.4 µg/L), and, like particulate fraction microcystins that year, did not follow a recognizable trend. It is, therefore, difficult to interpret the meaning of the higher percentage of dissolved microcystins that year, particularly because the regulation of toxin production in cyanobacteria is not well understood and the microcystin content of individual cells may be highly variable. However, toxins are released into the water column primarily following death and senescence of a toxigenic bloom, so it is likely that higher dissolved microcystin concentrations in 2008 represent longer or more frequent periods of decline in the microcystin-producing population. In 2009, dissolved microcystin concentrations were less than 0.2 µg/L through July 14. After July 14, the concentrations at all sites except site RPT increased over the next 3 weeks to their seasonal maximum concentrations (fig. 2); concentrations did not peak at site RPT until September 8. Through the remaining sampling period, concentrations at all sites were highly variable. Maximum concentrations in samples from site MDN were 1.44 and 3.93 µg/L measured on September 22, 2008, and August 3, 2009, respectively. Samples from site WMR collected on June 30, 2008, and August 3, 2009, contained maximum concentrations of 0.48 and 2.49 µg/L, respectively.

Table 2. Median, range, and percentage of values below the method quantitation limit of microcystin concentrations in different size fractions collected between July 7 and September 14, Upper Klamath Lake, Oregon, 2007–09.

[Microcystin values less than 0.1 µg/L (not censored) were detected, but appear to be less than the method quantitation limit after calculating aqueous concentrations from the total extracted biomass and sample volumes. The summary percent of total toxins each size fraction represents does not include 2007 data. Dissolved microcystin concentrations were not measured in 2007. **Abbreviations:** MQL, method quantitation limit; >, greater than; <, less than; µm, micrometer; µg/L, micrograms per liter; µg/mg, micrograms per milligram; –, samples not analyzed (no data); NA, not applicable]

Analyte	2007				2008					2009				
	Number of samples	Median (µg/L)	Range (µg/L)	<MQL (percent)	Number of samples	Median (µg/L)	Range (µg/L)	<MQL (percent)	Median percentage of total	Number of samples	Median (µg/L)	Range (µg/L)	<MQL (percent)	Median percentage of total
Dissolved (µg/L)	0	–	–	–	20	0.27	0.18–1.10	0	85.8	67	1.49	<MQL–4.87	15	43
Microcystins														
Small (1.5–63 µm) (µg/L)	18	<MQL	<MQL–0.04	56	20	0.03	<MQL–0.10	35	1.5	67	<MQL	<MQL–0.10	57	0
Large (>63 µm) (µg/L)	18	1.87	0.44–17.4	0	20	0.20	0.01–1.55	0	11.7	67	1.37	0.001–24.4	0	56
Large (>63 µm) (µg/mg)	18	0.09	0.03–1.42	0	20	0.02	0.002–0.36	0	NA	67	0.15	0.001–0.96	0	NA

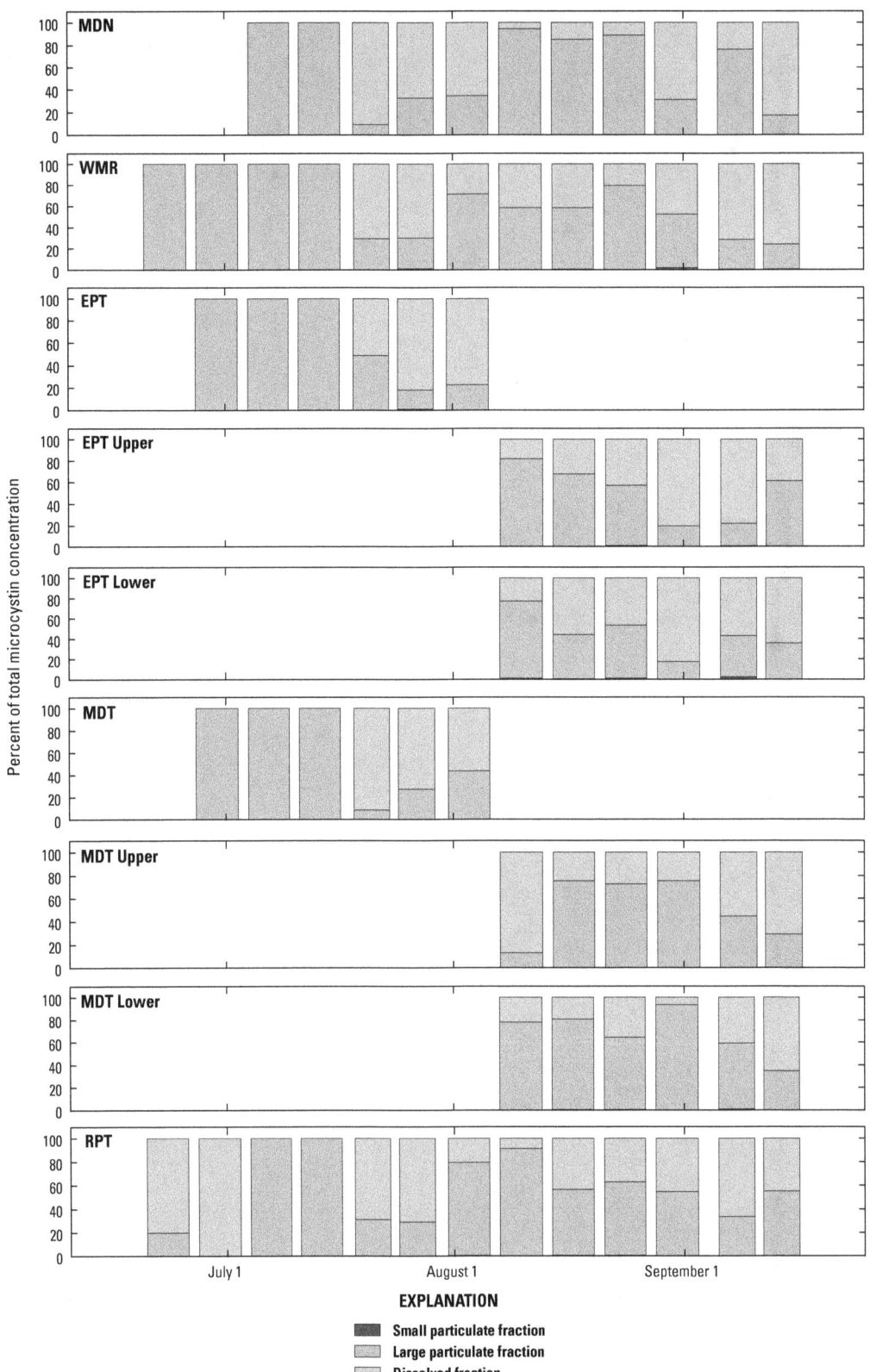

Figure 5. Percentage of microcystin concentrations representing each size fraction at sites in Upper Klamath Lake, Oregon, 2009.

Table 3. Median and range of large (>63 μm) particulate (cell associated) and dissolved microcystin concentrations at each sample site between July 7 and September 14, Upper Klamath Lake, Oregon, 2007–09.

[Depth-integrated samples from sites EPT and MDT were collected between July 7 and August 3, 2009. Samples collected at these sites on and after August 10, 2009, were taken at one-quarter (upper) and three-quarter (lower) depths in the water column. The numbers of depth-integrated samples are indicated by italics. All other samples were collected at discrete depths. Dissolved microcystin concentrations were not measured in 2007. Only sites MDN and WMR were sampled in all years of the study. **Abbreviations:** MQL, method quantitation limit; μm, micrometer; μg/L, microgram per liter; <, less than; >, greater than; –, no samples collected]

Sample site	2007			2008			2009			Summary		
	Number of samples	Median (μg/L)	Range (μg/L)	Number of samples	Median (μg/L)	Range (μg/L)	Number of samples	Median (μg/L)	Range (μg/L)	Number of samples	Median (μg/L)	Range (μg/L)
Large particulate (cell associated) microcystin concentrations (>63 μm)												
MDN	3	0.73	0.64–17.4	5	0.03	0.01–0.33	11	1.01	0.01–11.8	19	0.47	0.01–17.4
WMR	3	6.41	0.64–7.35	5	0.53	0.19–1.13	11	0.64	0.15–6.23	19	0.64	0.15–7.35
EPT	3	1.92	0.44–12.6	0	–	–	5	0.20	0.001–0.78			
EPT Upper	0	–	–	0	–	–	6	2.85	0.24–6.54			
EPT Lower	0	–	–	0	–	–	6	1.83	0.31–2.39			
MDT	3	1.48	1.33–12.5	0	–	–	5	0.09	0.03–3.82			
MDT Upper	0	–	–	0	–	–	6	3.98	0.51–9.81			
MDT Lower	0	–	–	0	–	–	6	4.80	1.26–24.4			
MDL	0	–	–	5	0.23	0.02–1.55	0	–	–			
RPT	3	1.81	1.32–10.1	0	–	–	11	1.37	0.04–9.06			
NBI	0	–	–	5	0.03	0.01–0.29	0	–	–			
HDB	3	2.30	1.22–13.1	0	–	–	0	–	–			
Dissolved microcystin concentrations												
MDN				5	0.23	0.21–1.10	11	1.05	<MQL–3.93	16	0.61	<MQL–3.93
WMR				5	0.25	0.18–0.36	11	0.85	<MQL–2.49	16	0.47	<MQL–2.49
EPT				0	–	–	5	0.33	<MQL–2.64			
EPT Upper				0	–	–	6	1.95	1.01–3.12			
EPT Lower				0	–	–	6	2.26	0.60–2.98			
MDT				0	–	–	5	0.35	<MQL–4.87			
MDT Upper				0	–	–	6	2.42	2.28–3.38			
MDT Lower				0	–	–	6	2.46	1.43–3.19			
MDL				5	0.49	0.21–0.75	0	–	–			
RPT				0	–	–	11	1.02	<MQL–2.77			
NBI				5	0.35	0.20–0.47	0	–	–			
HDB				0	–	–	0	–	–			

Relations Between Microcystin Concentrations and Other Water Quality Variables

Correlation analysis was performed to identify potentially important associations between microcystin concentrations measured from 2007 to 2009 and environmental factors that have been previously associated with microcystin occurrences in other systems and with the growth and decline of *A. flos-aquae* (these factors may similarly influence growth and decline of toxigenic *M. aeruginosa*) in Upper Klamath Lake. Laboratory studies of batch culture incubations, for instance, show that high concentrations of nitrogen and phosphorus in freshwaters may favor the growth of toxic *Microcystis* strains over nontoxic ones (Vézie and others, 2002). In addition, many experimental studies have shown that intracellular microcystin content or *M. aeruginosa* growth is favored by elevated concentrations of phosphorus accompanied by low nitrogen to phosphorus ratios (Schindler, 1977; Kotak and others, 2000; Xie and others, 2003; Rantala and others, 2006; Chen and others, 2009); increases in total or dissolved inorganic nitrogen (Te and Gin, 2011); increases in nitrogen and phosphorus (Vézie and others, 2002; Wilhelm and others, 2011); higher water temperature and pH (Dokulil and Teubner, 2000; Jacoby and others, 2000; Te and Gin, 2011); or dissolved oxygen concentrations (Te and Gin, 2011). In the current study, correlation analysis was performed with total microcystin concentrations and with the dissolved and large (> 63 µm) particulate fractions separately to determine if changes in tested parameters correlated with microcystins during the different stages of a potential toxigenic *M. aeruginosa* bloom cycle; the presence of microcystins in the particulate fraction was used to represent a *M. aeruginosa* bloom period, and higher dissolved microcystin concentrations were assumed to occur as the bloom declined (analogous to using chlorophyll *a* as a biomass indicator for *A. flos-aquae* in this system). Linear correlations of high significance are extremely difficult to achieve when the important connections may be more tied to patterns of succession or sequences of events. Therefore, it should be emphasized that correlation analysis was used in this study to identify potential environmental influences on microcystin occurrence (based on results of previous studies) and to support data trends used to form hypotheses concerning the sequence of events required to promote growth of toxigenic cells and the associated increase in microcystin concentrations in Upper Klamath Lake. Among the variables tested, total microcystins (data collected in 2008 and 2009) correlated significantly (*p* < 0.05) with DIN:DIP ratio; p-values in correlations between water temperature and water column stability (RTRM) measured at site MDN also were near significance at *p* < 0.1 (table 4). Dissolved microcystins measured in 2008 and 2009 correlated significantly with only water column stability at site MDN, and more variables, TN:TP ratio, DIP, DIN:DIP ratio, pH, and water column stability at site MDN significantly correlated with large particulate (cell associated) microcystin concentrations measured in all years. The negative correlation between large particulate microcystins and DIN:DIP ratio was the most statistically significant relation found in the analysis, and p-values less than 0.1 also were found between microcystins in the large particulate fraction and total phosphorus and dissolved oxygen concentrations. The low p-values obtained from the positive correlations of microcystins in the large particulate fraction with total phosphorus and DIP, along with the significant negative correlations with TN:TP and DIN:DIP ratios (table 4) and increasing concentrations of total phosphorus and DIP over each sample season (fig. 2), suggest a possible association between phosphorus availability and the presence of cell associated microcystins (in the large particulate fraction) at sites sampled in all years (MDN and WMR). Water column stability measured at site MDN also correlated significantly or nearly significantly with all forms of microcystins measured, illustrating the tendency for toxigenic cells to produce intracellular microcystins under low mixing (stable water column) conditions.

Table 4. Spearman rank order correlations (r) between microcystin concentrations and other environmental variables, Upper Klamath Lake, Oregon, 2007–09.

[Data from only sites MDN and WMR were used in the analysis. Correlations with total and dissolved microcystin concentrations include only 2008 and 2009 data. Dissolved microcystin concentrations were not measured in 2007. Chlorophyll *a* data collected in 2007 were of poor quality and not reported. **Boldface type** indicates significant correlation at $p < 0.05$; * indicates $p < 0.1$. **Abbreviations:** TN:TP, the ratio of total nitrogen to total phosphorus; DIP, dissolved inorganic phosphorus; DIN, dissolved inorganic nitrogen; DIN:DIP, the ratio of dissolved inorganic nitrogen to dissolved inorganic phosphorus; <, less than]

Environmental variable	Dissolved microcystins			Large particulate (cell associated) microcystins			Total microcystins		
	Number of samples	Spearman rank order correlation coefficient (r)	p-value (p)	Number of samples	Spearman rank order correlation coefficient (r)	p-value (p)	Number of samples	Spearman rank order correlation coefficient (r)	p-value (p)
Chlorophyll *a*	24	0.19	0.36	24	0.09	0.66	24	0.17	0.42
Total nitrogen	24	0.19	0.35	32	0.11	0.56	24	0.03	0.90
Total phosphorus	24	0.18	0.38	32	0.30	0.09*	24	0.13	0.53
TN:TP	24	−0.19	0.37	32	−0.52	**0.01**	24	−0.31	0.14
DIP	24	0.16	0.45	32	0.43	**0.01**	24	0.18	0.39
DIN	24	0.04	0.86	32	−0.14	0.45	24	−0.15	0.49
DIN:DIP	24	−0.23	0.28	32	−0.61	**<0.01**	24	−0.43	**0.04**
Water temperature	24	−0.27	0.19	32	−0.05	0.77	24	−0.38	0.07*
pH	24	0.27	0.19	32	0.40	**0.02**	24	0.33	0.12
Dissolved oxygen concentration	24	0.16	0.45	32	0.30	0.09*	24	0.33	0.12
Wind speed	24	−0.18	0.40	32	−0.08	0.64	24	−0.22	0.30
Water column stability at site MDN	24	0.53	**0.01**	32	0.38	**0.03**	24	0.39	0.06*

Seasonal Variability, 2009

Analysis of median concentrations of total microcystins, chlorophyll *a*, total nutrients, and dissolved nutrients during the 2009 sample season reveals potentially important seasonal patterns among these parameters (fig. 6). Only data collected in 2009 are shown because total microcystin concentrations measured in 2008 were significantly lower than in 2007 or 2009 and because concentrations measured in 2007 do not included the dissolved fraction. Correlation analysis was performed between total microcystins or microcystins in the dissolved or large (> 63 µm) particulate fractions measured in 2009 and the suite of environmental variables used for the interannual correlation analysis (table 5).

Chlorophyll *a*

In 2009, the first lakewide maxima in chlorophyll *a* concentrations occurred while total microcystin concentrations were lowest (fig. 6A). Total microcystin concentrations began to increase in mid-July near the end of the *A. flos-aquae*-dominated bloom decline, then increased more rapidly during the bloom recovery, beginning July 21 and 27. A corresponding increase in *M. aeruginosa* cell density also was recorded at sites near those used in the current study (J. Kann, Aquatic Ecosystem Sciences, LLC, unpub. data, 2011). Between August and mid-September, total microcystin concentrations followed a similar pattern to chlorophyll *a* concentrations, peaking on August 17 and declining thereafter until the end of the season. As with results of the interannual correlation analysis, no correlation was found between chlorophyll *a* and large particulate (cell associated) microcystin concentrations in 2009 (table 5), which supports the low detection of microcystin concentrations during the first bloom and the observed high concentrations during the second bloom.

Total and Dissolved Nutrients

As seen in previous years in Upper Klamath Lake, total nitrogen and total phosphorus increased between May and October 2009 (fig. 6B). At the onset of both *A. flos-aquae*-dominated blooms (during late-June and late-August; fig. 6A), total nitrogen increased rapidly with chlorophyll *a*, suggesting that major bloom expansions during these periods primarily were associated with atmospheric nitrogen sequestration by nitrogen fixation. Benthic fluxes also add nitrogen to the water column (Kuwabara and others, 2007), as do riverine inputs, but these sources are smaller by comparison (Kann and Walker, 1999). Total phosphorus increased more steadily

than did total nitrogen during the 2009 season (fig. 6B), which suggests that internal loading from lake sediments, the most important source of phosphorus to this system, is less of an episodic phenomenon. Results of weekly sampling in 2009 also show that microcystin concentrations were highest after the mid-season increase in total phosphorus and nitrogen following the major bloom decline, and that, although the concentration of dissolved inorganic nitrogen decreased after July 27, concentrations were still sufficient to promote growth of non-diazotrophic, toxigenic strains of *M. aeruginosa*. Total phosphorus correlated positively with total microcystins ($r = 0.61$, $p = 0.02$), dissolved microcystins ($r = 0.69$, $p = 0.01$) and with microcystins in the large particulate fraction ($r = 0.63$, $p = 0.01$; table 5), and no correlations were found between microcystins and total nitrogen. This suggests a possible relation between microcystin and total phosphorus concentrations in 2009. It is not possible with the available data to determine whether the presence of phosphorus directly influences microcystin occurrence or if both factors are related to some other common variable, particularly because the correlation between large particulate microcystins and total phosphorus in all years combined was not significant at $p < 0.05$. However, it can be hypothesized that the positive correlation with total phosphorus concentrations in 2009 is due to an indirect association between microcystin occurrence (or *M. aeruginosa* growth) and phosphorus availability, because bioavailable phosphorus appears to regulate the *A. flos-aquae*-dominated bloom and given the apparent dependence of *M. aeruginosa* growth on nitrogen-fixation by *A. flos-aquae*.

Coincident with minimum concentrations of chlorophyll *a* near the end of July, concentrations of dissolved inorganic nitrogen (DIN) and dissolved inorganic phosphorus (DIP) reached a seasonal maximum as a result of nutrient release during cell lysis and decomposition of senesced phytoplankton (figs. 6C and 6D). These sharp increases and peaks in DIN and DIP concentrations occurred 2 weeks prior to the start of the increase in total microcystin concentrations on July 20 and before the peak in total microcystin concentrations was observed on August 10 (fig. 2). This suggests that the increase in available nitrogen and phosphorus promoted growth of both *A. flos-aquae* (as indicated by increasing chlorophyll *a*) and toxigenic *M. aeruginosa* (as indicated by increased microcystin concentrations). Diazotrophic species, such as *A. flos-aquae*, use DIN preferentially as a nitrogen source over nitrogen-fixation, which is a far more energy-consuming process, so it is likely that these organisms use N_2 after DIN is depleted. Dissolved microcystins correlated positively with DIP ($r = 0.53$, $p = 0.05$) and negatively with DIN:DIP ratio ($r = -0.55$, $p = 0.04$), but, as with total nitrogen, no correlations were observed between DIN and microcystins in any size fraction.

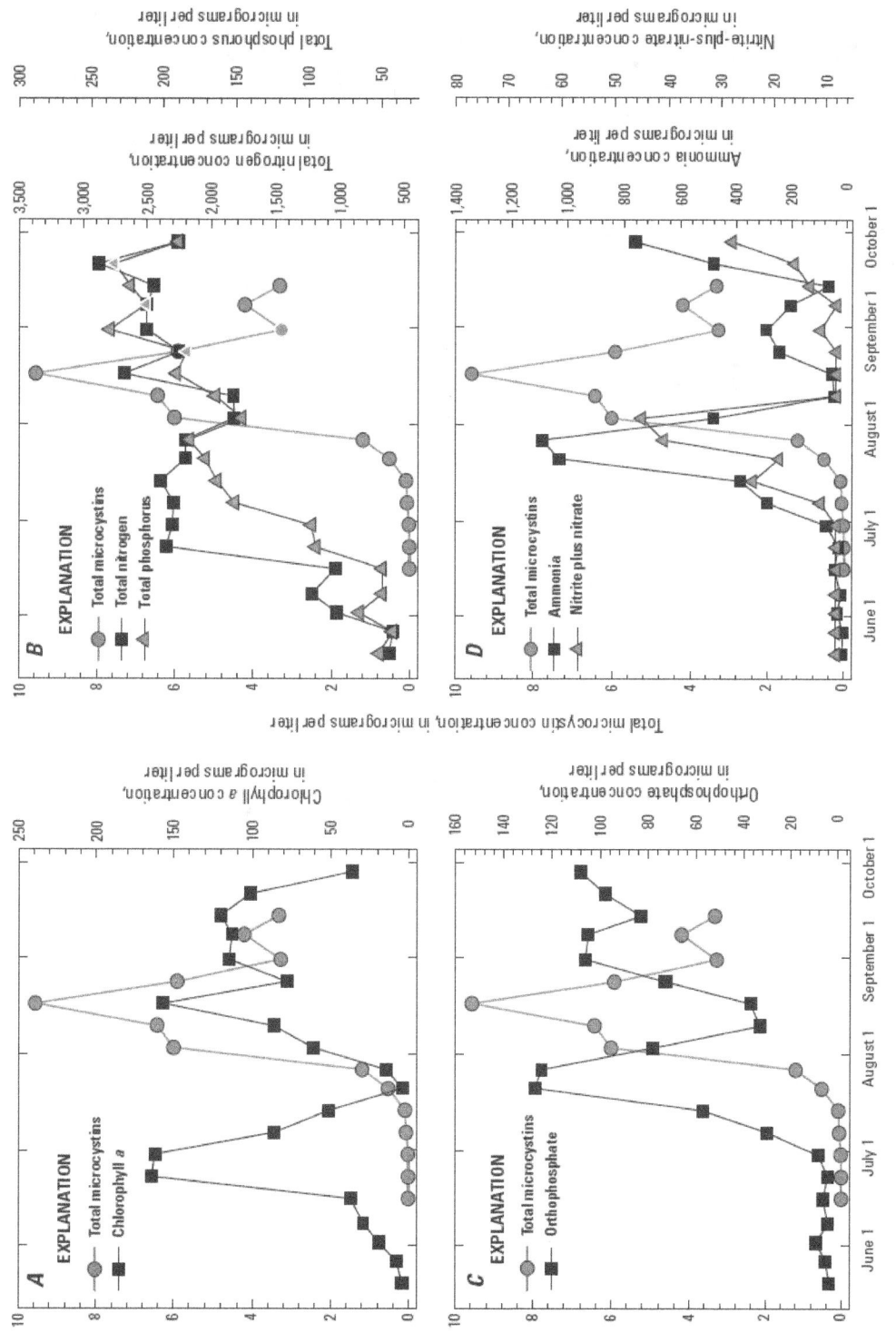

Figure 6. Median concentrations of total microcystins and (A) chlorophyll a, (B) total phosphorus and total nitrogen, (C) orthophosphate (dissolved inorganic phosphorus), and (D) ammonia and nitrite plus nitrate (dissolved inorganic nitrogen) in Upper Klamath Lake, Oregon, 2009.

Table 5. Spearman rank order correlations (r) between microcystin concentrations and other environmental variables, Upper Klamath Lake, Oregon, 2009.

[Correlations were determined between microcystin concentration and wind speed recorded at site MDL. Water column stability was determined by calculating the relative thermal resistance to mixing (RTRM; see text for explanation) at site MDN on cyanotoxin sample dates. **Boldface type** indicates significant correlation at $p < 0.05$; *, indicates $p < 0.1$. **Abbreviations:** TN:TP, the ratio of total nitrogen to total phosphorus; DIN, dissolved inorganic nitrogen; DIP, dissolved inorganic phosphorus; DIN:DIP, the ratio of dissolved inorganic nitrogen to dissolved inorganic phosphorus; TPN, total particulate nitrogen; TPP, total particulate phosphorus; TPC, total particulate carbon; TPN:TPP, the ratio of total particulate nitrogen to total particulate phosphorus; <, less than]

Environmental variable	Dissolved microcystins			Large particulate (cell associated) microcystins			Total microcystins		
	Number of samples	Spearman rank order correlation coefficient (r)	p-value (p)	Number of samples	Spearman rank order correlation coefficient (r)	p-value (p)	Number of samples	Spearman rank order correlation coefficient (r)	p-value (p)
Chlorophyll a	14	0.12	0.68	14	0.05	0.84	14	0.05	0.84
Total nitrogen	14	0.30	0.29	14	0.19	0.51	14	0.16	0.56
Total phosphorus	14	0.69	**0.01**	14	0.63	**0.01**	14	0.61	**0.02**
TN:TP	14	−0.73	**<0.01**	14	−0.69	**0.01**	14	−0.69	**0.01**
DIN	14	0.18	0.53	14	0.13	0.64	14	0.13	0.64
DIP	14	0.53	**0.05**	14	0.41	0.13	14	0.41	0.14
DIN:DIP	14	−0.55	**0.04**	14	−0.71	**<0.01**	14	−0.70	**<0.01**
TPN	12	0.17	0.59	12	0.14	0.65	12	0.11	0.72
TPP	12	0.33	0.28	12	0.46	0.12	12	0.44	0.14
TPC	12	0.14	0.65	12	0.13	0.68	12	0.09	0.77
TPN:TPP	12	−0.46	0.12	12	−0.60	**0.04**	12	−0.64	**0.02**
Water temperature	14	−0.08	0.77	14	0.01	0.98	14	0.01	0.95
pH	14	0.33	0.24	14	0.38	0.17	14	0.32	0.26
Dissolved oxygen	14	0.15	0.59	14	0.23	0.42	14	0.23	0.42
Wind speed	14	−0.34	0.23	14	−0.44	0.11	14	−0.48	0.08*
Water column stability at site MDN	14	0.61	**0.02**	14	0.46	0.09*	14	0.51	**0.05**

Nutrient Ratios

The late-July 2009 increase in total microcystin concentrations coincided with low TN:TP ratios and the second *A. flos-aquae*-dominated bloom; TN:TP ratios were higher during the first bloom than when this second bloom occurred (figs. 6A and 7A). Concentrations of total nitrogen and total phosphorus increased over the season (fig. 6B), so the decrease in TN:TP between the first and second blooms was due to a greater increase in total phosphorus, relative to total nitrogen concentrations, possibly as a consequence of decreased nitrogen-fixation during the *A. flos-aquae*-bloom decline. TN:TP ratios correlated negatively with total, dissolved, and particulate (cell associated) microcystin concentrations (total microcystins: $r = -0.69$, $p = 0.01$; dissolved microcystins: $r = -0.73$, $p = < 0.01$; large particulate microcystins: $r = -0.69$, $p = 0.01$); in correlation analysis using data from all years, only large particulate microcystins were significantly correlated. This result is consistent with other studies (Jacoby and others, 2000; Kotak and others, 2000; Xie and others, 2003), but does not support interspecific competition for nitrogen as the primary reason for diazotrophic *A. flos-aquae* dominance (by biovolume) over non-diazotrophic *M. aeruginosa* in Upper Klamath Lake, at least during mid- to late-summer.

Microcystin concentrations began to increase during the period of higher DIN:DIP ratios (prior to August 1; fig. 7B) that resulted from the release of nutrients during the first major bloom decline. Concentrations of dissolved microcystins initially peaked along with DIN:DIP (after August 1), but the peak in particulate fraction microcystins was not observed until after the sharp decrease in DIN:DIP, which occurred as nutrient uptake increased during the second *A. flos-aquae*-dominated bloom (figs. 2 and 7B). Therefore, it appears that intracellular toxin production was not adversely affected by the low DIN:DIP ratios (and the accompanying low ammonia concentrations, fig. 6D) after August 10. Total, dissolved, and large particulate (cell associated) microcystins were negatively correlated with DIN:DIP ratios (total microcystins: $r = -0.70$, $p < 0.01$; large particulate microcystins: $r = -0.71$, $p < 0.01$; dissolved microcystins: $r = -0.55$, $p = 0.04$), in agreement with results of the interannual correlation analysis and with the observed decrease in DIN:DIP prior to the peak in particulate microcystin concentrations.

Particulate Carbon, Nitrogen, and Phosphorus

Changes in median concentrations of total particulate carbon (TPC), total particulate nitrogen (TPN), and total particulate phosphorus (TPP), which represent changes in the population density of primarily *A. flos-aquae* over time in Upper Klamath Lake, followed patterns similar to that of chlorophyll *a* (figs. 6A and 8) in 2009, increasing during the first bloom in late June-early July, decreasing during the bloom decline, and increasing again, along with total microcystins, as the second bloom developed in mid-August. As with chlorophyll *a*, the relatively low median particulate nutrient concentrations in the last half of July corresponded with peaks in median concentrations of dissolved nutrients (figs. 6C and 6D), in that particulate nutrients transitioned into dissolved nutrients during the bloom decline. However, no correlation was found between total particulate nutrient and microcystin concentrations (table 5).

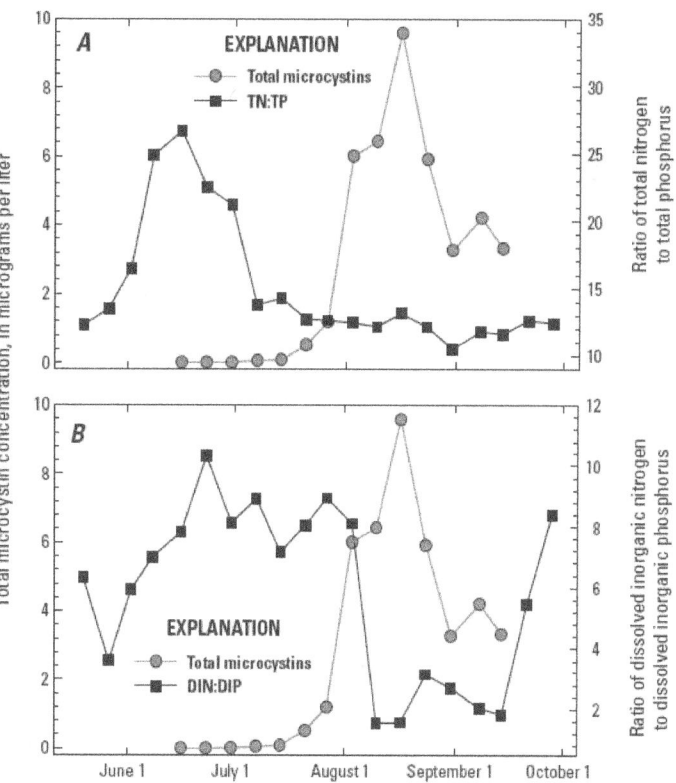

Figure 7. Median concentrations of total microcystins and (*A*) the ratio of total nitrogen to total phosphorus and (*B*) the ratio of dissolved inorganic nitrogen to dissolved inorganic phosphorus in Upper Klamath Lake, Oregon, 2009.

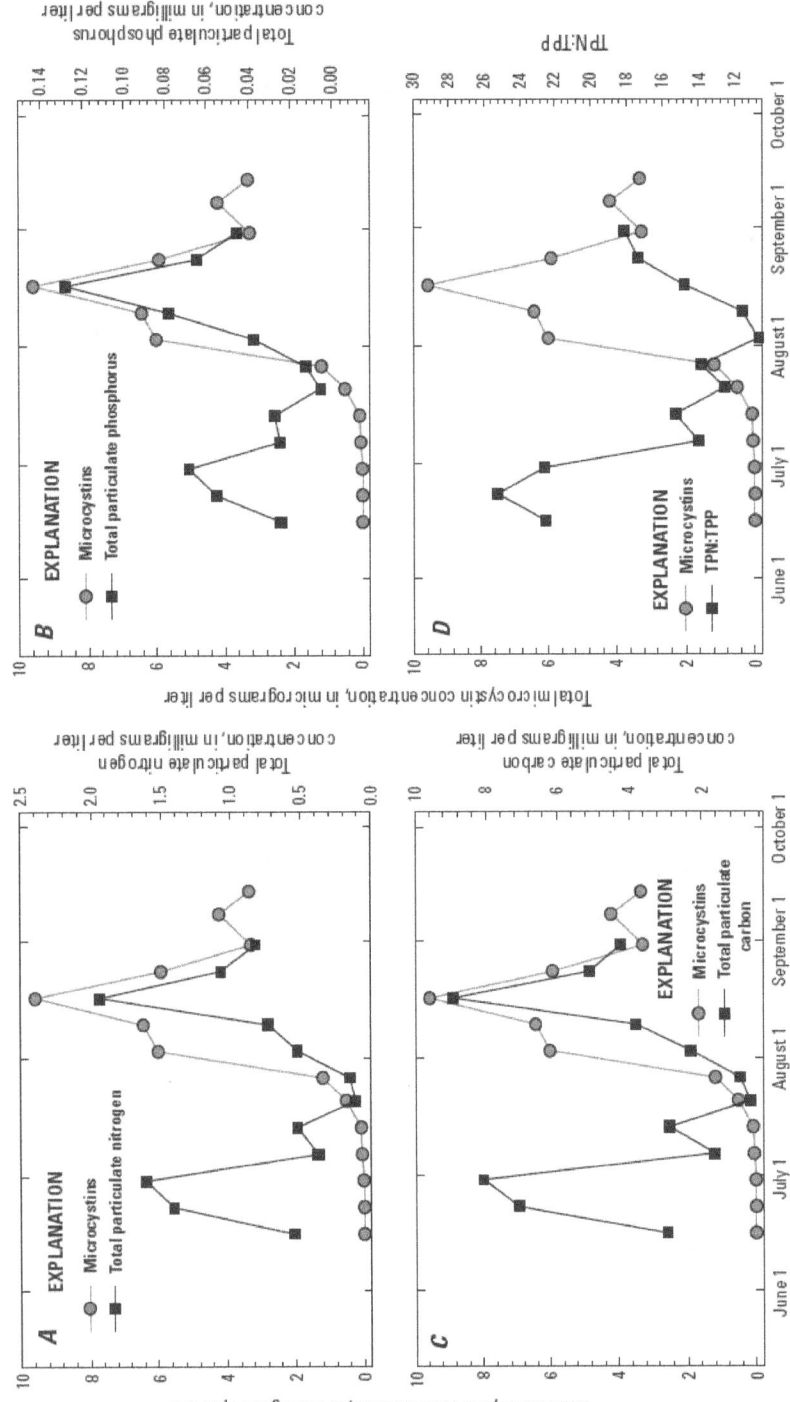

Figure 8. Median concentrations of total microcystins and (A) total particulate nitrogen, (B) total particulate phosphorus, (C) total particulate carbon, and (D) the ratio of total particulate nitrogen to total particulate phosphorus in Upper Klamath Lake, Oregon, 2009.

Median ratios of TPN:TPP decreased between early July and early August similarly to the pattern observed for the ratio of total nitrogen to total phosphorus (figs. 7A and 8D). Median TPN:TPP ratios of 24–25 were observed along with peak chlorophyll *a* concentrations during the first *A. flos-aquae* bloom between June 16 and 23, whereas the median TPN:TPP ratio was closer to 14 during the second peak in the *A. flos-aquae* bloom on August 17. Concentrations of particulate inorganic phosphorus (PIP) were 29 ±12 percent of TPP and exhibited no clear seasonal trend (data not shown), which indicates that ratios of organic PN:PP were closer to 34 or 35 during the first bloom and near 15 during the second bloom. Although the organic particulate material was not composed exclusively of *A. flos-aquae*, the dominance of *A. flos-aquae* in phytoplankton biomass of Upper Klamath Lake indicates a lower overall N:P ratio in cells comprising the second *A. flos-aquae*-dominated bloom, which is consistent with a switch from nutrient-limited to nutrient-replete conditions (Klausmeier and others, 2004). The most rapid increase and seasonal maxima of total microcystin concentrations (and *M. aeruginosa* cell densities at adjacent sample sites; J. Kann, Aquatic Ecosystem Sciences, LLC, unpub. data, 2011) occurred with the lowest seasonal TPN:TPP ratios, which accounts for the significant correlation between total particulate nutrient ratios and total and large particulate (cell associated) microcystin concentrations.

Continuous Monitor and Meteorological Variables

In 2009, total and dissolved microcystin concentrations were positively correlated with water column stability measured at site MDN (total: $r = 0.51$, $p = 0.05$; dissolved: $r = 0.61$, $p = 0.02$; the p-value was low but not significant at $p < 0.05$ between large particulate microcystins and water column stability: $r = 0.46$, $p = 0.09$). Total microcystin concentrations also correlated negatively near significance ($p < 0.1$) with daily median wind speed ($r = -0.48$, $p = 0.08$), which, together with the positive correlation between microcystin concentrations and water column stability, agree with results of the interannual correlation analysis that shows accumulation of toxigenic *M. aeruginosa* cells, as with *A. flos-aquae* colonies, occurs in a stable water column. Elevated pH and dissolved oxygen concentrations are characteristic of dense phytoplankton growth, and the non-significant correlation observed between water column pH, dissolved oxygen concentrations, and total or particulate fraction microcystin concentrations may indicate that the increased productivity during bloom development and the large decreases in pH and dissolved oxygen concentrations between the first and second bloom periods (during bloom decline) do not influence microcystin occurrence in the lake.

Spatial Variability of Microcystin Concentrations, 2009

Spatial patterns of large particulate (cell associated) and dissolved microcystins in Upper Klamath Lake directly reflect wind-driven current flow. Prevailing westerly to northwesterly winds create a dominantly clockwise circulation pattern in the lake that results in southward flow that is broad and shallow along the eastern shoreline and narrow flow northward through the deepest part of the lake (the trench) along the western shoreline (Wood and others, 2006, 2008). Previous work also has shown that the heaviest accumulation of phytoplankton (and the highest degree of cell senescence during periods of bloom decline) occurs in the deeper areas along the trench (Hoilman and others, 2008), which is the likely cause of high dissolved and particulate fraction microcystin concentrations measured at site MDT (table 3) in 2009. The highest concentrations of particulate nutrients and chlorophyll *a* also were measured at site MDT that year (chlorophyll *a* also was highest at site MDT in 2008; chlorophyll *a* data were not reported in 2007; Kannarr and others, 2010), which further indicates that the highest accumulation of buoyant colonies occurs along the western shoreline. Similar results have been reported in other studies of Upper Klamath Lake (Jassby and Kann, 2010; Kann, 2010). Northward flow through this area appears to have transported particulates to sites WMR and EPT, where relatively moderate to high concentrations of large particulate (cell associated) microcystins were measured. Likewise, shallow flow southward through the middle and eastern portion of the lake may have contributed to the occurrence of moderate dissolved and higher particulate microcystin concentrations at site RPT, the southernmost site (fig. 9).

Between August 10 and September 14, 2009, samples for microcystin analysis were collected at discrete depths (one-quarter and three-quarters depths in the water column) at the deep sites, MDT and EPT, in order to determine whether microcystin concentrations varied in a consistent way with depth. However, the results of ANOVA and two-sample t-tests indicated no significant differences between microcystin concentrations (any size fraction) in samples from the upper and lower portions of the water column. In addition, no significant relation was found in correlation analyses between the microcystin concentrations (any size fraction) collected at the upper or lower depths and with nutrient (dissolved or particulate) concentrations collected at the same time and at the same depths (results not shown).

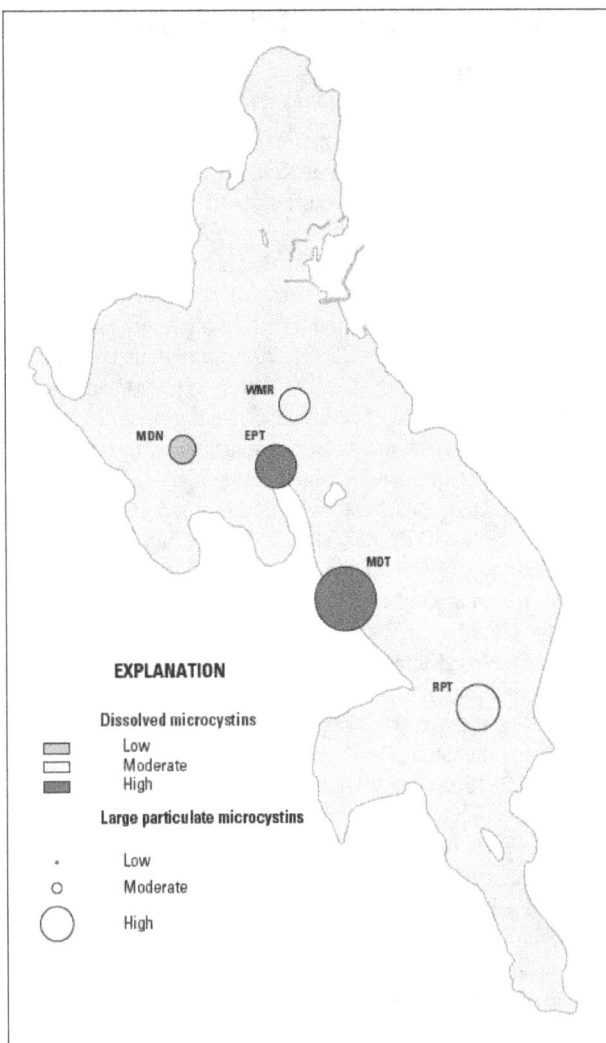

Figure 9. Relative median concentrations of microcystins in the dissolved and large (> 63 μm) particulate (cell associated) fractions at sites in Upper Klamath Lake, Oregon, 2009.

Relation Between *Microcystis aeruginosa* and *Aphanizomenon flos-aquae*

Unlike *A. flos-aquae*, *M. aeruginosa* does not fix nitrogen, which, at times, appears to be a growth-limiting nutrient for phytoplankton in Upper Klamath Lake, given the dominance of diazotrophic cyanobacteria, such as *A. flos-aquae* (Reuter and others, 1993). However, members of the *Microcystis* genus are highly adaptable and frequently become the dominant organism in eutrophic systems, including reservoirs of the Klamath River (Jacoby and Kann, 2007; Moisansder and others, 2009) and other Oregon lakes (Barbiero and Kann, 1994). *Microcystis* sp. have a high

affinity to absorb dissolved inorganic nitrogen (Takamura and others, 1987), can store large amounts of intracellular phosphorus as polyphosphate (Jacobson and Halmann, 1982), and can take up phosphorus directly from attached bacteria (Jiang and others, 2007). *Microcystis* sp. cells also can tolerate strong irradiance (Paerl and others, 1985), overwinter as vegetative cells (without sporulating; Preston and others, 1980), and, like *A. flos-aquae*, they can adjust their buoyancy to occupy the best position for receiving optimum light intensity for photosynthesis (Ibelings and others, 1991). Such characteristics contribute to the success of *Microcystis* sp. in temperate and tropical climates, but the ability of this microorganism to coexist in direct competition with *A. flos-aquae* is unknown because there have been few, if any, direct comparisons between their critical physiological parameters (Yamamoto, 2009). Recent *in situ* nutrient enrichment experiments in the Copco and Iron Gate Reservoirs on the Klamath River, California, showed that, during the summer, nitrogen was frequently the primary nutrient limiting growth of *M. aeruginosa* and microcystin concentration, although changes in per-cell toxin content or the ratio in abundances of toxic versus nontoxic strains could have contributed to the observed trends (Kann and Corum, 2009; Bozarth and others, 2010). The effects of nitrogen addition were clearest when biomass and overall microcystin concentrations were lowest, and, on several occasions, secondary phosphorus limitation was observed (cell abundance increased when phosphorus was added in combination with nitrogen; Moisander and others, 2009).

The ecological relation between *M. aeruginosa* and *A. flos-aquae,* whether it is competitive, facilitative, or neutral, has implications for the management of Upper Klamath Lake. If the relation is competitive, the occurrence of toxigenic *M. aeruginosa* may increase if nutrient management (reduction of phosphorus inputs, the focus of most proposed plans) successfully limits the *A. flos-aquae* bloom. However, if the relation is facilitative, an overall decrease or elimination of *A. flos-aquae* also may eliminate *M. aeruginosa*, given that this species already occurs in low abundance here. These alternatives can not be evaluated definitively with the data collected in the current study, and the relation between *A. flos-aquae* and *M. aeruginosa* may be more complex and variable throughout the season, but the results of this work do permit the creation of testable hypotheses, primarily the hypothesis that toxigenic *M. aeruginosa* and microcystin occurrence are associated with the second of two *A. flos-aquae*-dominated blooms observed in most years. The most notable difference in lake conditions between 2007 and 2009 was the timing of the dominant bloom cycle and severity of the first bloom decline. In years with the highest microcystin concentrations, the first bloom declined sharply in July (mid-July in 2007 and late July in 2009) and was followed by a second bloom about 2 weeks later. In contrast, only one bloom was observed in 2008, which declined later and did not culminate until about the third week in August (Kann, 2010). This resulted in the highest concentrations of DIP and DIN occurring relatively late that year and, together with the patterns in dissolved nutrient concentrations and ratios observed other years, leads to an additional hypothesis that toxigenic *M. aeruginosa* growth and (or) microcystin occurrence is stimulated directly by the release of DIN during the major *A. flos-aquae*-dominated bloom decline but is dependent, overall, on the presence of phosphorus to regulate growth and decline of *A. flos-aquae*.

In Upper Klamath Lake, nitrogen fixation by *A. flos-aquae* early in the season, when concentrations of *A. flos-aquae* typically are highest, may facilitate the growth of toxigenic *M. aeruginosa* after the first major bloom decline by supplying new nitrogen to the system as *A. flos-aquae* cells lyse and decompose; this appears to be a stronger relation in years, such as 2009, with a well-defined, lakewide bloom cycle. However, because the *A. flos-aquae*-dominated bloom cycle appears to be regulated more by changes in phosphorus availability (given that *A. flos-aquae* is able to fix N_2 when DIN is not available), support of *M. aeruginosa* growth by an increase in new nitrogen favors phosphorus availability indirectly as a more important factor overall for regulation of microcystin concentrations. This also helps to explain the observed seasonal patterns in microcystin and chlorophyll *a* concentrations and why correlations were significant between microcystin and total or dissolved phosphorus concentrations in 2009 and not significant between microcystin and nitrogen (TN or DIN) concentrations. The results of this study show that high concentrations of DIN and DIP followed the major lakewide bloom decline in 2009 and may have promoted growth of toxigenic *M. aeruginosa* (as indicated by the occurrence of microcystins) in concert with the development of a second large *A. flos-aquae* bloom. These patterns were also observed in previous analyses of cyanobacterial blooms in Clear Lake, California, where summer release of ammonia from the decomposition of the spring *A. flos-aquae* bloom and the presence of naturally abundant phosphorus stimulated *Microcystis* growth and recovery of the dominant bloom former, *A. flos-aquae* (Horne, 1975; Horne and Goldman, 1994). Similar to results of the Clear Lake study, the microcystin data presented here do not indicate growth of toxigenic cells during the first *A. flos-aquae* bloom. Results of monthly sampling collected in 2007 do not provide the same temporal resolution as 2009 data, but the highest microcystin concentrations measured that year occurred on August 1, about 2 weeks after the bloom decline and 2 weeks earlier than the highest concentrations measured in 2009. Results of alternate-week sampling in 2008 showed uniformly low concentrations relative to the other 2 years, which probably is related to the absence of a large, early *A. flos-aquae*-dominated bloom and the late (and less severe) bloom decline observed that year.

In 2009, *M. aeruginosa* began to increase with the major decline in the *A. flos-aquae* bloom, was at very low concentrations prior to that decline, and continued to increase rapidly during the second *A. flos-aquae* bloom that followed the decline. Therefore, growth of toxigenic *M. aeruginosa* may be favored by the decrease in *A. flos-aquae* during the bloom decline (if these species compete directly) and by the

increased availability of nutrients (both DIN and DIP) during this time, but the occurrence of toxigenic *M. aeruginosa* does not seem to be adversely affected by the return of the *A. flos-aquae* bloom later in the season. *M. aeruginosa* appears to continue growing and (or) producing microcystins while co-existing with *A. flos-aquae* during the second bloom. As such, the relation between these species appears to have shifted in the latter half of the season from being based on competition to being more neutral, although other factors, independent of resource competition with *A. flos-aquae*, may have kept microcystin concentrations low earlier in the season. However, the difference in nitrogen to phosphorus ratios (TN:TP and TPN:TPP) measured between the first and second *A. flos-aquae* blooms should be noted, in that they indicate changes in nitrogen-to-phosphorus stoichiometry in phytoplankton cells under different environmental conditions. The determinants of optimal cellular nitrogen:phosphorus stoichiometry under different ecological scenarios have been previously derived and modeled (Klausmeier and others, 2004) and, in Upper Klamath Lake, indicate that the higher nitrogen to phosphorus ratios during the first *A. flos-aquae* bloom may have resulted from the allocation of more nutrients to phosphorus-poor cellular resource-acquisition machinery (for photosynthesis and nitrogen fixation) under competitive equilibrium. Likewise, the lower nitrogen to phosphorus ratios characteristic of the second *A. flos-aquae* bloom may have resulted from cells using more nutrients for phosphorus-rich protein assembly machinery (ribosomes) during exponential growth under nutrient replete conditions. If so, it is plausible that *A. flos-aquae* may have lower physiological requirements for nitrogen than toxigenic *M. aeruginosa* and, therefore, may "share" more nitrogen during the second bloom. A facilitative relation between diazotrophic and non-diazotrophic groups, in which the presence of the diazotrophs supported a larger population of non-diazotrophs than possible in the absence of the diazotroph, was recently observed in a marine environment (Agawin and others, 2007). The degree to which the relation between *M. aeruginosa* and *A. flos-aquae* is facilitative or neutral may be an important factor for determining the effects of using nutrient reduction to diminish or eliminate the *A. flos-aquae* bloom on the presence of *M. aeruginosa* and microcystins in Upper Klamath Lake.

Implications for Juvenile Sucker Health

Between 2007 and 2009, microcystin concentrations in the large (> 63 μm) particulate (cell assoicated) fraction measured in water column samples were elevated, which suggests a potential threat of microcystin exposure to fish and other wildlife ingesting particulates in Upper Klamath Lake. Poor water quality (extremely low dissolved oxygen concentrations, high pH, and elevated ammonia concentrations) has been proposed as a contributing factor for occasional periods of high mortality and overall population decline in Lost River and shortnose suckers (Martin and Saiki, 1999; Saiki and others, 1999). However, sharp decreases in age-0 catch rates between August and September over the last 5 years (Bottcher and Burdick, 2010), combined with low catches of age-1 and older suckers (Hendrixson and others, 2007; Terwilliger and others, 2008) indicate that juvenile suckers suffer widespread, episodic mortality during the first 2 years of life. The mechanism for this mortality is not known, but as many as one-half of juvenile suckers collected from Upper Klamath Lake in 2007 (VanderKooi and others, 2010) and 14 percent of juveniles collected from the central and northern regions of the lake in 2009 (n = 36; C. Ottinger, U.S. Geological Survey, unpub. data, 2010) exhibited liver or kidney damage consistent with microcystin ingestion. Furthermore, the disappearance of juvenile suckers from trap net catches has occurred 2 to 3 weeks after peak concentrations of microcystins have been measured (Burdick and others, 2009; Bottcher and Burdick, 2010).

Observational field studies and tissue analyses have documented the effects of microcystin exposure in whitefish (Ernst and others, 2001), tilapia (Magalhaes and others, 2001), and flounder (Sipia and others, 2001), but few studies have evaluated the primary mechanisms of cyanotoxin exposure in natural environments, oral ingestion (by feeding), and passage of dissolved toxins across gill membranes (Malbrouck and Kestemont, 2006). Recently, a preliminary risk assessment of shortnose and Lost River sucker microcystin exposure was conducted to determine the daily dry weight dosage per kilogram of live fish weight (Malbrouck and Kestemont, 2006) based on microcystin concentrations measured in 2007 (when the highest concentrations of particulate fraction microcystins were observed). If these fish consume 1 percent of their body weight per day of (dried) toxigenic cyanobacteria producing the most toxic microcystin isomer, microcystin-LR (these factors are currently unknown), at the peak microcystin concentration observed in this study (1.42 µg/mg dry weight; measured in 2007), a fish weighing 1 kg would receive a dose of 14,150 (µg/kg)/d (K. Echols, U.S. Geological Survey, written commun., 2007). This value is approximately 25 times the LC50 intra-peritoneal dose (the concentration required for 50 percent mortality) of microcystin-LR (550 µg/kg) for carp (Rabergh and others, 1991) and trout (Tencalla and others, 1994). Juvenile suckers (and larvae) may be more susceptible to toxins than adults (fish weighing closer to 1 kg) because of their larger surface area-to-volume ratios, higher metabolic rates, and the vulnerability of their key developmental processes. However, more work is needed to understand the dietary intake of toxigenic species in these fish, the concentration (or concentration range) of ingested microcystins that damage liver tissue and promote mortality in juvenile suckers, and if sufficient microcystin concentrations to cause the tissue damage or mortality rates observed are present within the materials these fish consume in Upper Klamath Lake.

Summary and Conclusions

As an extension of the long-term water quality monitoring program on Upper Klamath Lake and to support the U.S. Geological Survey's effort to determine the causes of juvenile Lost River and shortnose sucker population decline, water samples were collected between 2007 and 2009 and analyzed for intracellular (particulate) and extracellular (dissolved) microcystins and cylindrospermopsins. Measurements using enzyme-linked immunosorbent assays (ELISA) showed that microcystin concentrations at sampled sites were much higher than cylindrospermopsins, which were not present in most samples analyzed. In all years of the study, microcystins occurred primarily in dissolved and large (> 63 µm) particulate forms, rather than in the small (1.5–63 µm) particulate form and at concentrations that may pose a threat to fish and other wildlife. Concentrations of large particulate microcystins expressed volumetrically were highest in 2007, when total nutrient and orthophosphate concentrations were highest, and lowest in 2008. However, when concentrations of large particulate microcystins are expressed gravimetrically, the median concentration was highest in 2009. Samples collected in 2009 contained higher and more variable concentrations of dissolved microcystins. However, most of the 2008 samples contained microcystins primarily in the dissolved fraction although, in 2009, dissolved microcystins comprised less than one-half of the total concentrations measured that year. Between July and September in 2007 and 2009, samples from sites MDT and HDB contained the highest concentrations of large particulate microcystins and exhibited the widest range of values (samples from site HDB were collected only in 2007 and were not analyzed for dissolved microcystins). Between 2008 and 2009, dissolved microcystins were more concentrated in samples from sites MDT, EPT, and MDN (based on median and peak concentrations). Concentrations also were relatively high at MDL, but this site was sampled only in 2008 when concentrations were lower, overall.

In 2009, the first lakewide *Aphanizomenon flos-aquae*-dominated bloom, as indicated by chlorophyll *a* concentrations, occurred while microcystin concentrations generally were lowest. However, after the bloom recovered, microcystin concentrations followed a similar pattern to those of chlorophyll *a*. Microcystin concentrations (daily median values) also increased with total nitrogen and total phosphorus concentrations, which, in turn, increased rapidly with chlorophyll *a* during the second bloom of the season. Peak concentrations of dissolved inorganic nitrogen (DIN) and dissolved inorganic phosphorus (DIP) accompanied the minimum in chlorophyll *a* near the end of July and decreased as total microcystin concentrations began to increase. Ratios of total nitrogen to total phosphorus (TN:TP) decreased sharply, relative to early-season conditions, during the second *A. flos-aquae*-dominated bloom and as microcystin concentrations increased. Early increases in microcystin concentrations coincided with seasonally high DIN:DIP ratios, but particulate microcystin concentrations continued to rise after a sharp decrease in DIN:DIP occurred, suggesting that intracellular microcystin occurrence was not adversely affected by low DIN:DIP ratios or by the accompanying low ammonia concentrations.

Changes in median concentrations of total particulate carbon, total particulate nitrogen, and total particulate phosphorus over time followed a pattern similar to that of chlorophyll *a* and exhibited sharp decreases that corresponded with peaks in median concentrations of dissolved nutrients. Median total particulate nitrogen to total particulate phosphorus (TPN:TPP) ratios were higher during the first *A. flos-aquae* bloom than during the second bloom peak in mid-August when total microcystins increased most rapidly.

Understanding the ecological interactions between *M. aeruginosa* and *A. flos-aquae* in Upper Klamath Lake is important for implementing effective lake management, in that activities (including the limitation of phosphorus inputs), which may affect the growth and abundance of *A. flos-aquae* will likely influence the presence of toxigenic *Microcystis aeruginosa*. Early in the sample season, nitrogen fixation by the abundant *A. flos-aquae* population appears to provide new nitrogen to facilitate growth of toxigenic cells (and other non-diazotrophs), which suggests that, because phosphorus plays a major role in regulating the dominant bloom cycle, limited phosphorus availability may be indirectly important for regulating growth of these groups. Later in the season, these species appear to co-exist mutually, rather than compete for resources, and the lower nitrogen to phosphorus ratios (TN:TP and TPN:TPP) measured during this time indicate that, on the cellular level, more nutrients may be allocated to exponential growth rather than to resource-acquisition machinery, which is more heavily supported under competitive equilibrium early in the season when nutrient concentrations are low.

Results of this study contribute to understanding the spatiotemporal dynamics of microcystin occurrence in Upper Klamath Lake and the possible environmental influences on microcystin concentrations or changes in the biomass of microcystin producers. Analyses of data collected between 2007 and 2009 have led to the formation of testable hypotheses that will become the focus of future monitoring and research. In particular, growth of toxigenic *M. aeruginosa* and elevated microcystin concentrations appear to be linked with the second *A. flos-aquae*-dominated bloom in years with a well-defined cycle of growth and decline (two bloom periods typically are observed) as a result of the large increase in dissolved nutrients released from decomposing cells. In addition, although toxigenic *M. aeruginosa* growth and (or) microcystin production are likely dependent on the release of DIN following the first major *A. flos-aquae*-dominated bloom decline (*A. flos-aquae* is diazotrophic and *M. aeruginosa* is not), time-series and correlation analyses suggest that the occurrence of microcystins in Upper Klamath Lake may be strongly regulated, indirectly, by phosphorus availability,

which, in turn, drives the seasonal *A. flos-aquae*-dominated bloom cycle. Furthermore, ongoing and future collaborative work with fishery biologists, chemists, and specialists in other disciplines is needed to determine if the endangered suckers in Upper Klamath Lake consume toxigenic cyanobacteria through the food chain and if these fish consume microcystins at concentrations necessary to promote significant tissue damage and mortality. Knowledge of the environmental, physiological, and ecological factors that regulate microcystin production in Upper Klamath Lake is critical for effective lake management, minimization of health risks to wildlife, livestock, and humans, and for understanding the risk of toxin exposure to endangered populations of native fish species inhabiting the lake.

Acknowledgments

We thank the three peer reviewers of this article for their helpful comments. We also gratefully acknowledge the assistance of Scott VanderKooi and the personnel of the U.S. Geological Survey (USGS) Klamath Falls Field Station for facilitating the water quality field program with the use of boats, trucks, field equipment, and office and laboratory space. Many people contributed to the fieldwork for this study, and we also acknowledge their efforts: Mary Lindenberg, Kristofor Kannarr, Dean Snyder, Lee Simons, Matthew Abel, William Wright, Amari Dolan-Caret, Elena Ceballos, William Ayers, Cynthia King, Kristin Harbin, D. Blake Eldridge, and Matthew Wilson from the USGS Klamath Falls Field Station; Amy Brooks, Micelis Doyle, and Matt Johnston from the USGS Oregon Water Science Center. We also thank and acknowledge the laboratory work of Kevin Feltz, Lynne Johnson, and Jesse Arms at the USGS Columbia Environmental Research Center.

References Cited

Agawin, N.S.R., Rabouille, S., Veldhuis, M.J.W., Servatius, L., Hol, S., van Overzee, H.M.J., and Huisman, J., 2007, Competition and facilitation between unicellular nitrogen-fixing cyanobacteria and non-nitrogen-fixing phytoplankton species: Limnology and Oceanography, v. 52, p. 2233–2248.

Aranda-Rodriguez, R., Tillmanns, A., Benoit, F.M., Pick, F.R., Harvie, J., and Solenaia, L., 2005, Pressurized liquid extraction of toxins from cyanobacterial cells: Environmental Toxicology, v. 20, p. 390–396.

Aspila, I., Agemian, H., and Chau, A.S.Y., 1976, A semi-automated method for the determination of inorganic, organic, and total phosphate in sediments: Analyst, v. 101, p. 187–197.

Barbiero, R.P., and Kann, J., 1994, The importance of benthic recruitment to the population development of *Aphanizomenon flos-aquae* and internal loading in a shallow lake: Journal of Plankton Research, v. 16, p. 1581–1588.

Bortelson, G.C., and Fretwell, M.O., 1993, A review of possible causes of nutrient enrichment and decline of endangered sucker populations in Upper Klamath Lake, Oregon: U.S. Geological Survey Water-Resources Investigations Report 93–4087, 24 p.

Bottcher, J.L., and Burdick, S.M., 2010, Temporal and spatial distribution of endangered juvenile Lost River and shortnose suckers in relation to environmental variables in Upper Klamath Lake, Oregon—2009 annual data summary: U.S. Geological Survey Open-File Report 2010–1261, 42 p. (Also available at http://pubs.usgs.gov/of/2010/1261/.)

Bozarth, C.S., Schwartz, A.D., Shepardson, J.W., Colwell, F.S., and Dreher, T.W., 2010, Population turnover in a *Microcystis* bloom results in predominantly nontoxigenic variants late in the season: Applied and Environmental Microbiology, v. 76, no. 15, p. 5207–5213.

Bradbury, J.P., Colman, S.M., and Reynolds, R.L., 2004, The history of recent limnological changes and human impact on Upper Klamath Lake, Oregon: Journal of Paleolimnology, v. 31, p. 151–165.

Buettner, M., and Scoppettone, G.G., 1990, Life history and status of catostomids in Upper Klamath Lake, Oregon—Completion report: U.S. Fish and Wildlife Service, National Fisheries Research Center, Reno Field Station, Nevada, 119 p.

Burdick, S.M., VanderKooi, S.P., and Anderson, G.O., 2009, Spring and summer spatial distribution of endangered juvenile Lost River and shortnose suckers in relation to environmental variables in Upper Klamath Lake, Oregon—2007 Annual Report: U.S. Geological Survey Open-File Report 2009–1043, 56 p. (Also available at http://pubs.usgs.gov/of/2009/1043/.)

Bureau of Reclamation, 2000, Klamath Project—Historical operation: Klamath Falls, Oregon, Bureau of Reclamation, 53 p. plus appendices.

Carey, C.C., Haney, J.F., and Cottingham, K.L., 2007, First report of microcystin-LR in the cyanobacterium *Gloeotrichia echinulata*: Environmental Toxicology, v. 22, no. 3, p. 337–339.

Carmichael, W.W., 1994, The toxins of cyanobacteria: Scientific American, v. 270, p. 78–86.

Carmichael, W.W., 1997, The cyanotoxins, *in* Callow, J.A.. ed., Advances in Botanical Research, v. 37, p. 211–256.

Carmichael, W.W., 2000, Harvesting of *Aphanizomenon flos-aquae* Ralfs ex Born & Flah. Var. *flos- aquae* (Cyanobacteria) from Klamath Lake for human dietary use: Journal of Applied Phycology, v. 12, p. 585–595.

Carmichael, W.W., Pinotti, M.H., and Fraleigh, P.C., 1986, Toxicity of a clonal isolate of the cyanobacterium (blue-green alga) *Microcystis aeruginosa* from Lake Erie: The Ohio Journal of Science, v. 86, p. 53–153.

Chen, H., Burke, J.M., Mosindy, T., Fedorak, P.M., and Prepas, E.E., 2009, Cyanobacteria and microcystin-LR in a complex lake system representing a range of trophic status—Lake of the Woods, Ontario, Canada: Journal of Plankton Research, v. 31, no. 9, p. 993–1008.

Dokulil, M.T., and Teubner, K., 2000, Cyanobacterial dominance in lakes: Hydrobiologia, v. 438, p. 1–12.

Downing, J.A., Watson, S.B., and McCauley, E., 2001, Predicting cyanobacteria dominance in lakes: Canadian Journal of Fisheries and Aquatic Sciences, v. 58, p. 1905–1908.

Downing, T.G., Meyer, C., Gehringer, M.M., and van de Venter, M., 2005, Microcystin content of *Microcystis aeruginosa* is modulated by nitrogen uptake rate relative to specific growth rate or carbon fixation rate: Environmental Toxicology, v. 20, p. 257–262.

Eilers, J., Kann, J., Cornett, J., Moser, K., St. Amand, A., and Gubala, C., 2004, Recent paleolimnology of Upper Klamath Lake, Oregon: Hydrobiologie, v. 520, p. 7–18.

Erdner, D.L., Dyble, J., Parsons, M.L., Stevens, R.C., Hubbard, K.A., Wrabel, M.L., Moore, S.K., Lefebvre, K.A., Anderson, D.M., Bienfang, P., Bidigare, R.R., Parker, M.S., Moeller, P., Brand, L.E., and Trainer, V.L., 2008, Centers for oceans and human health—A unified approach to the challenge of harmful algal blooms: Environmental Health 7 (Suppl. 2), p. S2.

Ernst, B., Hitzfeld, B., and Dietrich, D., 2001, Presence of *Planktothrix* sp. and cyanobacterial toxins in Lake Ammersee, Germany and their impact on whitefish (*Coregonus lavaretus* L.): Environmental Toxicology, v. 16, p. 483–488.

Falconer, I.R., 1999, An overview of problems caused by toxic blue-green algae (cyanobacteria) in drinking and recreational water: Environmental Toxicology, v. 14, p. 5–12.

Forsberg, C., and Ryding, S.O., 1980, Eutrophication parameters and trophic state indices in 30 Swedish waste-receiving lakes: Archiv für Hydrobiologie, v. 89, p. 189–207.

Gilroy, D.J., Kauffman, K.W., Hall, R.A., Huang, X., and Chu, F.S., 2000, Assessing potential health risks from microcystin toxins in blue-green algae dietary supplements: Environmental Health Perspective, v. 108, no. 5, p. 435–439.

Gleit, A., 1985, Estimation for small normal data sets with detection limits: Environmental Science and Technology, v. 19, no. 12, p. 1201–1206.

Graham, J.L., and Jones, J.R., 2007, Microcystin distribution in physical size class separations of natural plankton communities: Lake and Reservoir Management, v. 23, p. 161–168.

Graham, J.L., Jones, J.R., Jones, S.B., Downing, J.A., and Clevenger, T.E., 2004, Environmental factors influencing microcystin distribution and concentration in the Midwestern United States: Water Research, v. 38, p. 4395–4404.

Graham, J.L., Loftin, K.A., Ziegler, A.C., and Meyer, M.T., 2008, Guidelines for design and sampling for cyanobacterial toxin and taste-and-odor studies in lakes and reservoirs: U.S. Geological Survey Scientific Investigations Report 2008–5038, 39 p. (Also available at http://pubs.usgs.gov/sir/2008/5038/.)

Graham, J.L., Loftin, K.A., and Kamman, N., 2009, Monitoring recreational freshwaters: LakeLine, v. 29, p. 16–22.

Harada, K.I., Ogawa, K., Kimura, Y., Murata, H., Suzuki, M., Thron, P.M., Evans, W.R., and Carmichael, W., 1991, Microcystins from *Anabaena flos-aquae* NRC 525-17: Chemical Research in Toxicology, v. 4, p. 535–540.

Hendrixson, H.A., Burdick, S.M., and VanderKooi, S.P., 2007, Near-shore and offshore habitat use by endangered, juvenile Lost River and shortnose suckers in Upper Klamath Lake, Oregon—Annual report 2004: Report of U.S. Geological Survey, Western Fisheries Research Center, Klamath Falls Field Station to Bureau of Reclamation, Mid-Pacific Region, Klamath Falls, Oregon.

Hoilman, G.R., Lindenberg, M.K., and Wood, T.M., 2008, Water quality conditions in Upper Klamath and Agency Lakes, Oregon, 2005: U.S. Geological Survey Scientific Investigations Report 2008–5026, 44 p. (Also available at http://pubs.usgs.gov/sir/2008/5026/.)

Horne, A.J., 1975, The ecology of Clear Lake phytoplankton: Special report of the Clear Lake Algal Research Unit, Lakeport, California, 116 p.

Horne, A.J. and Goldman, C.R., 1994, Limnology: New York, McGraw Hill, 576 p.

Hughes, E.O., Gorham, P.R., and Zehnder, A., 1958, Toxicity of a unialgal culture of *Microcystis aeruginosa*: Canadian Journal of Microbiology, v. 4, p. 225–236.

Ibelings, B.W., Mur, L.R., and Walsby, A.E., 1991, Diurnal changes in buoyancy and vertical distribution in populations of *Microcystis* in two shallow lakes: Journal of Plankton Research, v. 13, p. 419–436.

Jacobson, L., and Halmann, M., 1982, Polyphosphate metabolism in the blue-green alga *Microcystis aeruginosa*: Journal of Plankton Research, v. 4, p. 481–488.

Jacoby, J.M., and Kann, J., 2007, The occurrence and response to toxic cyanobacteria in the Pacific Northwest, North America: Lake and Reservoir Management, v. 23, p. 123–143.

Jacoby, J.M., Collier, D.C., Welch, E.B., Joan Hardy, F., and Crayton, M., 2000, Environmental factors associated with a toxic bloom of *Microcystis aeruginosa*: Canadian Journal of Fisheries and Aquatic Sciences, v. 57, p. 231–240.

Jassby, A., and Kann, J., 2010, Upper Klamath Lake monitoring program—Preliminary analysis of status and trends for 1990–2009: Prepared for Klamath Tribes Natural Resources Department, Chiloquin, Oregon, June 16, 2010.

Jiang, L., Yang, L., Xiao, L., Shi, X., Gao, G., and Qin, B., 2007, Quantitative studies on phosphorus transference occurring between *Microcystis aeruginosa* and its attached bacterium (*Pseudomonas* sp.): Hydrobiologia, v. 581, p. 161–165.

Johnson, D.M., 1985, Atlas of Oregon lakes: Corvallis, Oregon, Oregon State University Press, 317 p.

Jones, C.A., and Welch, E.B., 1990, Internal phosphorus loading related to mixing and dilution in a dendritic shallow prairie lake: Journal of the Water Pollution Control Federation, v. 62, p. 847–852.

Jones, S.B., and Jones, J.R., 2002, Seasonal variation in cyanobacterial toxin production in two Nepalese lakes: Verhandlungen des Internationalen Verein Limnologie, v. 28, p. 1017–1022.

Kann, J., 1997, Ecology and water quality dynamics of a shallow hypereutrophic lake dominated by cyanobacteria (*Aphanizomenon flos-aquae*): Chapel Hill, University of North Carolina, Ph.D. Dissertation, 110 p.

Kann, J., 2010, Upper Klamath Lake 2009 data summary report: Chiloquin, Oregon, Report to Klamath Tribes Natural Resources Department [variously paged].

Kann, J., and Corum, S., 2009, Toxigenic *Microcystis aeruginosa* bloom dynamics and cell density/chlorophyll *a* relationships with microcystin toxin in the Klamath River, 2005–2008: Technical Memorandum to the Karuk Tribe of California [variously paged].

Kann, J., and Corum, S., 2010, Middle Klamath River toxic cyanobacteria trends, 2009: Technical memorandum to the Karuk Tribe of California [variously paged].

Kann, J., and Smith, V.H., 1999, Chlorophyll as a predictor of elevated pH in a hypereutrophic lake—Estimating the probability of exceeding critical values for fish success using parametric and nonparametric models: Canadian Journal of Fisheries and Aquatic Sciences, v. 56, p. 2262-2270.

Kann, J., and Walker, W.W., 1999, Nutrient and hydrologic loading to Upper Klamath Lake, Oregon, 1991–1998: Klamath Falls, Oregon, Report to Klamath Tribes Natural Resources Department and Bureau of Reclamation [variously paged].

Kann, J., and Welch, E.B., 2005, Wind control on water quality in shallow, hypereutrophic Upper Klamath Lake, Oregon: Lake and Reservoir Management, v. 21, no. 2, p. 149–158.

Kannarr, K.E., Tanner, D.Q., Lindenberg, M.K., and Wood, T.M., 2010, Water-quality data from Upper Klamath and Agency Lakes, Oregon, 2007–08: U.S. Geological Survey Open-File Report 2010-1073, 28 p. (Also available at http://pubs.usgs.gov/of/2010/1073/.)

Klamath Tribes, 1994, Written testimony of the Klamath Tribes before the Water and Power Subcommittee of the U.S. Senate Committee on Energy and Natural Resources (Senators B. Bradley and M. Hatfield presiding): Klamath Falls, Oregon, [variously paged].

Klausmeier, C.A., Litchman, E., Daufresne, T., and Levin, S.A., 2004, Optimal nitrogen-to-phosphorus stoichiometry of phytoplankton: Nature, v. 429, p. 171–174.

Kotak, B.G., Lam, A., Prepas, E.E., and Hrudey, S.E., 2000, Role of chemical and physical variables in regulating microcystin-LR concentration in phytoplankton of eutrophic lakes: Canadian Journal of Fisheries and Aquatic Sciences, v. 57, p. 1584–1593.

Krishnamurthy, T., Carmichael, W.W., and Sarver, E.W., 1986, Toxic peptides from freshwater cyanobacteria (blue-green algae). I. Isolation, purification, and characterization of peptides from *Microcystis aeruginosa* and *Anabaena flos-aquae*: Toxicon, v. 24, p. 865–873.

Kurmayer, R., and Christiansen, G., 2009, The genetic basis of toxin production in cyanobacteria: Freshwater Reviews, v. 2, no. 1, p. 31–50.

Kurmayer, R., Christiansen, G., and Chorus, I., 2003, The abundance of microcystin-producing genotypes correlates positively with colony size in *Microcystis* sp. and determines its microcystin net production in Lake Wannsee: Applied and Environmental Microbiology, v. 69, no. 2, p. 787–795.

Kuwabara, J.S., Lynch, D.D., Topping, B.R., Murphy, F., Carter, J.L., Simon, N.S., Parchaso, F., Wood, T.M., Lindenberg, M.K., Wiese, K., and Avanzino, R.J., 2007, Quantifying the benthic source of dissolved nutrients to the water column of Upper Klamath Lake, Oregon: U.S. Geological Survey Open-File Report 2007-1276, 39 p. (Also available at http://pubs.usgs.gov/of/2007/1276/.)

Lawrence, J.F., Niedzwiadek, B., Menard, C., Lau, B.P.Y., Lewis, D., and Kuper-Goodman, T., 2001, Comparison of liquid chromatography/mass spectrometry, ELISA, and phosphatase assay for the determination of microcystins in blue-green algae products: Journal of the Association of Official Analytical Chemists International, v. 84, no. 4, p. 1035–1044.

Lindenberg, M.K., Hoilman, G.R., and Wood, T.M., 2009, Water quality conditions in Upper Klamath and Agency Lakes, Oregon, 2006: U.S. Geological Survey Scientific Investigations Report 2008–5201, 54 p. (Also available at http://pubs.usgs.gov/sir/2008/5201/.)

Magalhaes, V.F., Soares, R.M., and Azevedo, S.M.F.O., 2001, Microcystin contamination in fish from the Jacarepagua Lagoon (RJ, Brazil)—Ecological implication and human health risk: Toxicon, v. 39, p. 1077–1108.

Malbrouck, C., and Kestemont, P., 2006, Effects of microcystins on fish: Environmental Toxicology and Chemistry, v. 25, p. 72–86.

Martin, B.A., and Saiki, M.K., 1999, Effects of ambient water quality on the endangered Lost River sucker in Upper Klamath Lake, Oregon: Transactions of the American Fisheries Society, v. 128, p. 953–967.

Meriluoto, J.A.O., Sandstrom, A., Eriksson, J.E., Remaud, G., Craig, A.G., and Chattopadhyaya, J., 1989, Structure and toxicity of a peptide hepatotoxin from the cyanobacterium *Oscillatoria agardhii*: Toxicon, v. 27, p. 1021–1034.

Moisander, P.H., Lehman, P.W., Ochiai, M., and Corum, S., 2009, Diversity of *Microcystis aeruginosa* in the Klamath River and San Francisco Bay delta, California USA: Aquatic Microbial Ecology, v. 57, no. 1, p. 19–31.

National Research Council, 2004, Endangered and threatened fishes in the Klamath River basin: Washington, D.C., National Academy Press.

Orr, P.T., and Jones, G.J., 1998, Relationships between microcystin production and cell division rates in nitrogen-limited *Microcystis aeruginosa* cultures: Limnology and Oceanography, v. 43, p. 1604–1614.

Oudra, B., Loudiki, M., Vasconcelos, V., Sabour, B., Sbiyyaa, B., Oufdou, K., and Mezrioui, N., 2002, Detection and quantification of microcystins from cyanobacteria strains isolated from reservoirs and ponds in Morocco: Environmental Toxicology, v. 17, no. 1, p. 32–39.

Paerl, H.W., 1988, Nuisance phytoplankton blooms in coastal, estuarine, and inland waters: Limnology and Oceanography, v. 33, p. 823–847.

Paerl, H.W., Bland, P.T., Bowles, N.D., and Haibach, M.E., 1985, Adaptation to high-intensity, low- wavelength light among surface blooms of the cyanobacterium *Microcystis aeruginosa*: Applied and Environmental Microbiology, v. 49, p. 1046–1052.

Park, H.D., Iwami, C., Watanabe, M.F., Harada, K.I., Okino, T., and Hayashi, H., 1998, Temporal variabilities of the concentrations of intra- and extracellular microcystin and toxic *Microcystis* species in a hypertrophic lake, Lake Suwa, Japan (1991–1994): Environmental Toxicology and Water Quality, v. 13, p. 61–72.

Perkins, D., Kann, J., and Scoppettone, G.G., 2000, The role of poor water quality and fish kills in the decline of endangered Lost River and shortnose suckers in Upper Klamath Lake: U.S. Geological Survey Report to Bureau of Reclamation, Klamath Falls Project Office, Klamath Falls, Oregon, Contract 4-AA-29-12160.

Preston, T., Stewart, W.D.P., and Reynolds, C.S., 1980, Bloom-forming cyanobacterium *Microcystis aeruginosa* over-winters on sediment surface: Nature, v. 288, p. 365–367.

Preussel, K., Stüken, A., Wiedner, C., Chrous, I., and Fastner J., 2006, First report on cylindrospermopsin producing *Aphanizomenon flos-aquae* (Cyanobacteria) isolated from two German lakes: Toxicon, v. 47, p. 156–162.

Rabergh, C.M.I., Bylund, G., and Eriksson, J.E., 1991, Histopathological effects of MC-LR, a cyclic peptide toxin from the cyanobacterium *Microcystis aeruginosa* on common carp (*Cyprinus carpio* L.): Aquatic Toxicology, v. 20, p. 131–146.

Rantala, A., Rajaniemi-Wacklin, P., Lyra, C., Lepistö, L., Rintala, J., Mankiewicz-Boczek, J., and Sivonen, K., 2006, Detection of microcystin-producing cyanobacteria in Finnish lakes with genus-specific microcystin synthetase gene E (*mcyE*) PCR and associations with environmental factors: Applied and Environmental Microbiology, v. 72, no., 9, p. 6101–6110.

Reuter, J.E., Rhodes, C.L., Lebo, M.E., Kotzman, M., and Goldman, C.R., 1993, The importance of nitrogen in Pyramid Lake (Nevada, USA), a saline, desert lake: Hydrobiologia, v. 267, p. 179–189.

Reynolds, C.S., 1998, What factors influence the species composition of phytoplankton in lakes of different trophic status?: Hydrobiologia, v. 369/370, p. 11–26.

Saiki, M.K., Monda, D.P., and Bellerud, B.L., 1999, Lethal levels of selected water quality variables to larval and juvenile Lost River and shortnose suckers: Environmental Pollution, v. 105, p. 37–44.

Saker, M.L., Welker, M., and Vasconcelos, V.M., 2007, Multiplex PCR for the detection of toxigenic cyanobacteria in dietary supplements produced for human consumption: Applied Microbiology and Biotechnology, v. 73, p. 1136–1142.

Schindler, D.W., 1977, Evolution of phosphorus limitation in lakes: Science, v. 195, p. 260–262.

Scoppettone, G.C., and Vinyard, C.L., 1991, Life history and management of four lacustrine suckers, *in* Minckley, W.L., and Deacon, J.E., eds., Battle against extinction—Native fish management in the American west: The University of Arizona Press, Tucson, Arizona, p. 369–387.

Sedmak, B., and Elersek, T., 2006, Microcystins induce morphological and physiological changes in selected representative phytoplanktons: Microbial Ecology, v. 51, p. 508–515.

Sipia, V.O., Kankaanpaa, H.T., Flinkman J., Lahti, K., and Meriluoto, J.A.O., 2001, Time-dependent accumulation of cyanobacterial hepatotoxins in flounders (*Platichthys flesus*) and mussels (*Mytilus edulis*) from the Northern Baltic Sea: Environmental Toxicology, v. 16 p. 330–336.

Sivonen, K., and Jones, G., 1999, Cyanobacterial toxins, *in* Chorus, I., and Bartram, J., eds: Toxic cyanobacteria in water—A guide to public health significance, monitoring and management: London, E. & F.N. Spon, p. 41–111.

Smith, V., 1983, Low nitrogen to phosphorus ratios favor dominance of blue-green algae in lake phytoplankton: Science, v. 221, p. 669–671.

Takamura, N., Iwakuma, T., and Yasuno, M., 1987, Uptake of ^{13}C and ^{15}N (ammonium, nitrate and urea) by *Microcystis* in Lake Kasumigaura: Journal of Plankton Research, v. 9, p. 151–165.

Te, S.H., and Gin, K.Y.-H., 2011, The dynamics of cyanobacteria and microcystin production in a tropical reservoir of Singapore: Harmful Algae, v. 10, p. 319–329.

Tencalla, F., Dietrich, D., and Schlatter, C., 1994, Toxicity of *Microcystis aeruginosa* peptide toxin to yearling rainbow trout (*Onchorynchus mykiss*): Aquatic Toxicology, v. 30, p. 215–224.

Terwilliger, M.R., Simon, D.C., and Markle, D.F., 2008, Ecology of Upper Klamath Lake shortnose and Lost River suckers, 2006 Annual Report: Report of Oregon Cooperative Research Unit, Department of Fisheries and Wildlife, Oregon State University to U.S. Geological Survey, Corvallis, Oregon, and Klamath Project, Bureau of Reclamation, Klamath Falls, Oregon.

Trimbee, A.M., and Prepas, E.E., 1987, Evaluation of total phosphorus as a predictor of the relative biomass of blue-green algae with emphasis on Alberta lakes: Canadian Journal of Fisheries and Aquatic Sciences, v. 44, p. 1337-1342.

U.S. Environmental Protection Agency, 1997, Method No. 440.0 Determination of carbon and nitrogen in sediments and particulates of estuarine/coastal waters using elemental analysis: Cincinnati, Ohio, U.S. Environmental Protection Agency, 10 p.

U.S. Fish and Wildlife Service, 1993, Lost River (*Deltististes luxatus*) and shortnose (*Chasmistes brevirostris*) sucker recovery plan: Portland, Oregon, U.S. Fish and Wildlife Service.

U.S. Fish and Wildlife Service, 2002, Biological/conference opinion regarding the effects of operation of the U.S. Bureau of Reclamation project on the endangered Lost River sucker (*Deltististes luxatus*), shortnose sucker (*Chasmistes brevirostris*), threatened bald eagle (*Haliaeetus leucocephalus*) and proposed critical habitat for the Lost River/shortnose suckers for June 1, 2002–March 31, 2012: Klamath Falls, Oregon, U.S. Fish and Wildlife Service.

U.S. Geological Survey, variously dated, National field manual for the collection of water- quality data: U.S. Geological Survey Techniques of Water-Resources Investigations, book 9, chaps. A1–A5. (Also available at http://water.usgs.gov/owq/FieldManual/.)

VanderKooi, S.P., Burdick, S.M., Echols, K.R., Ottinger, C.A., Rosen, B.H., and Wood, T.M., 2010, Algal toxins in Upper Klamath Lake, Oregon—Linking water quality to juvenile sucker health: U.S. Geological Survey Fact Sheet 2009–3111, 2 p. (Also available at http://pubs.usgs.gov/fs/2009/3111/.)

Vézie, C., Rapala, J., Vaitomaa, J., Seitsonen, J., and Sivonen, K., 2002, Effect of nitrogen and phosphorus on growth of toxic and nontoxic *Microcystis* strains and on intracellular microcystin concentrations: Microbial Ecology, v. 43, p. 443–454.

Wagner, R.J., Boulger, R.W., Jr., Oblinger, C.J., and Smith, B.A., 2006, Guidelines and standard procedures for continuous water-quality monitors—Station operation, record computation, and data reporting: U.S. Geological Survey Techniques and Methods 1-D3, 51 p. plus attachments. (Also available at http://pubs.usgs.gov/tm/2006/tm1D3/.)

Welker, M., and Von Döhren, H., 2006, Cyanobacterial peptides—Nature's own combinational biosynthesis: FEMS Microbiology Reviews, v. 30, p. 530–563.

White, E., 1989, Utility of relationships between lake phosphorus and chlorophyll *a* as predictive tools in eutrophication control studies: New Zealand Journal of Marine and Freshwater Research, v. 23, no. 1, p. 35–41.

Wicks, R.J., and Theil, P.G., 1990, Environmental factors affecting the production of peptide toxins in floating scums of the cyanobacterium *Microcystis aeruginosa* in a hypertrophic African reservoir: Environmental Science and Technology, v. 24, p. 1413–1418.

Wiedner, C., Visser, P.M., Fastner, J., Metcalf, J.S., Codd, G.A., and Mur, L.R., 2003, Effects of light on the microcystin content of *Microcystis* strain PCC 7806: Applied and Environmental Microbiology, v. 69, no. 3, p. 1475–1481.

Wilhelm, S.W., Farnsley, S.E., LeCleir, G.R., Layton, A.C., Satchwell, M.F., DeBruyn, J.M., Boyer, G.L., Zhu, G., and Paerl, H.W., 2011, The relationships between nutrients, cyanobacterial toxins and the microbial community in Taihu (Lake Tai), China: Harmful Algae, v. 10, p. 207–215.

Williams, J.E., 1988, Endangered and threatened wildlife and plants—Determination of endangered status for the shortnose sucker and the Lost River sucker: Federal Register, v. 50, p. 27130–27134.

Wood, T.M., Hoilman, G.R., and Lindenberg, M.K., 2006, Water-quality conditions in Upper Klamath Lake, Oregon, 2002–04: U.S. Geological Survey Scientific Investigations Report 2006–5209, 52 p. (Also available at http://pubs.usgs.gov/sir/2006/5209/.)

Wood, T.M., Cheng, R.T., Gartner, J.W., Hoilman, G.R., Lindenberg, M.K., and Wellman, R.E., 2008, Modeling hydrodynamics and heat transport in Upper Klamath Lake, Oregon, and implications for water quality: U.S. Geological Survey Scientific Investigations Report 2008–5076, 48 p. (Also available at http://pubs.usgs.gov/sir/2008/5076/.)

World Health Organization, 2006, Guidelines for drinking-water quality; First addendum to third edition: *in* Recommendations, 3rd ed., v. 1, Geneva, World Health Organization, 515 p.

Xie, L., Xie, P., Li, S., Tang, H., and Liu, H., 2003, The low TN:TP ratio, a cause or result of *Microcystis* blooms?: Water Research, v. 37, p. 2073–2080.

Yamamoto, Y., 2009, Environmental factors that determine the occurrence and seasonal dynamics of *Aphanizomenon flos-aquae*: Journal of Limnology, v. 68, no. 1, p. 122–132.

Yu, S.H., 1989, Drinking water and primary liver cancer, *in* Tang, Z.Y., Wu, M.C., and Xia, S.S., eds., Primary Liver Cancer: New York, China Academic Publishers/Springer, p. 30–37.

Appendix A. Quality Control and Quality Assurance of Water Samples

In the present study, 17 percent of the 2007 microcystin samples and 20 percent of the 2008 and 2009 microcystin samples were collected for quality assurance. Each of the quality-assurance sample types, split samples and method replicate samples, were collected on alternate sample weeks in 2007 and 2009, but only method replicate samples were collected for microcystin analysis in 2008. Split samples were collected by dividing a single (composite) volume of lake water with a churn splitter, and were used to determine the variability in the laboratory method for microcystin analysis. Replicate samples were collected twice in rapid succession from the same location and analyzed to determine variability in the sample environment and analytical method. Quality-assurance samples were filtered, as were the primary samples, to create dissolved, small particulate (1.5–63 µm), and large particulate (> 63 µm) fractions, and the fractions were analyzed separately for microcystin concentrations. Therefore, the different fractions presented in table A1 within each year and each quality-assurance sample type are from the same environmental samples (for example, two split samples and three replicate samples were collected in 2007, and those samples were each divided into three fractions).

Results of quality-assurance sampling indicate that the variability in concentrations attributable to analysis and sampling methods was less than seasonal variability in each sample year (table A1). High median percent differences between the primary and quality-assurance (split or replicate) samples were generally observed when microcystin concentrations were near the method detection limit. This occurred with the small particulate fractions of quality-assurance samples collected in 2007 and 2008; in 2009, 71 percent of the small particulate fractions of quality-assurance samples contained microcystin concentrations less than the method detection limit, which is why table A1 shows 0 median percent difference in samples collected that year. Of the 3 years, samples collected in 2008 exhibited the highest variability within the dissolved and particulate fractions. The relative percent differences between primary and split or replicate samples of chlorophyll *a* and phaeophytin *a*

also were generally higher in 2008 than in 2009 or 2006 (chlorophyll *a* data collected in 2007 were not reported), in that more of the quality-assurance samples collected that year varied by 30 percent or more (Kannarr and others, 2010; Lindenberg and others, 2009). Similar results were obtained in quality-assurance analyses of total nutrient samples between 2006 and 2009. This suggests that much of the variability in microcystin quality-assurance samples collected in 2008 may be due to increased spatial variability (manifested as temporal variability at the sampling site) in the phytoplankton bloom that year and lower overall microcystin concentrations relative to 2007 and 2009.

Median percent differences in large particulate microcystin concentrations between the primary and quality-assurance samples were greater than 20 percent when measured as micrograms per liter in all years. However, with the exception of the 2009 split samples, when these concentrations were expressed on the basis of mass per dry weight of suspended solids, the relative percent differences were lower by more than 10 percent. These values were obtained from the same environmental samples, so the observed concentration variability may be a result of the method used for measuring microcystin concentrations within the solid phase of the water samples (that is, in measuring suspended material concentrations). In 2009, a high median percent difference was observed in the large particulate fraction of split samples reported as either micrograms per liter or micrograms per gram. Relative percent differences between split and primary samples were greater than 20 percent that year on July 21 (85 percent), July 27 (28 percent), and August 24 (27 percent) when the phytoplankton density was very low (as indicated by chlorophyll *a* concentrations and observations of field crew); microcystin concentrations were well above the detection limit. It is, therefore, likely that the concentration variability in these samples was due to the low volume of biomass collected.

Table A1 is available in a Microsoft© Excel workbook, which can be downloaded from http://pubs.usgs.gov/sir/2012/5069/.

Appendix B. Results of Microcystin Analysis

Table B1 is available in a Microsoft© Excel workbook, which can be downloaded from http://pubs.usgs.gov/sir/2012/5069/.